Lecture Notes in Mathematics   2226

More information about this series at http://www.springer.com/series/304

Gabriella Böhm

Hopf Algebras and Their Generalizations from a Category Theoretical Point of View

 Springer

Gabriella Böhm
Wigner Research Centre for Physics
Hungarian Academy of Sciences
Budapest, Hungary

ISSN 0075-8434   ISSN 1617-9692   (electronic)
Lecture Notes in Mathematics
ISBN 978-3-319-98136-9   ISBN 978-3-319-98137-6   (eBook)
https://doi.org/10.1007/978-3-319-98137-6

Library of Congress Control Number: 2018951456

Mathematics Subject Classification (2010): 16T10, 16T05, 18C15, 18D10, 18D15

This Springer imprint is published by the registered company Springer Nature Switzerland AG
The registered company address is: Gewerbestrasse 11, 6330 Cham, Switzerland

To Mom

Preface

These lecture notes were originally written for a short course delivered in March 2017 at the *Atlantic Algebra Centre of the Memorial University of Newfoundland, Canada*. It aimed to introduce to the audience—who had some familiarity with Hopf algebra theory—a category theoretical approach to the subject. Following this recent trend, Hopf algebras, as well as a wide variety of their generalizations, are regarded as opmonoidal monads on suitable monoidal categories. A main advantage of this unified treatment is a clear conceptual explanation of the behaviour of their modules.

It is a pleasure to thank the director of the *Atlantic Algebra Centre*, Yuri Bahturin, and Yorck Sommerhäuser for organizing this mini course, for their generous invitation, and for the warm hospitality experienced in St John's. Further thanks to the participants of the course for their interest, attention and helpful comments on the notes. The author is supported also by the Hungarian National Research, Development and Innovation Office NKFIH (grant K124138).

Budapest, Hungary   Gabriella Böhm
May 2018

The original version of the book was revised. The correction to the book is available at https://doi.org/10.1007/978-3-319-98137-6_9

Contents

Acronyms

alg	The category of algebras over a given field
bim(A)	The category of bimodules over a given algebra A
coalg	The category of coalgebras over a given field
disc(X)	The discrete category on the object set X
end(A)	The category of functors $\mathsf{A} \to \mathsf{A}$; i.e. nat(A, A)
hil	The category of complex Hilbert spaces (with continuous linear maps as morphisms)
ind(X)	The indiscrete category on the object set X
mod(A)	The category of modules over a given algebra A
nat(A, B)	The category of functors between given categories A, B (with natural transformations as morphisms)
set	The category of sets
span(X)	The category of spans over a given set X
vec	The category of vector spaces over a given field
vec^X	The category of vector spaces graded by a set X

Chapter 1
Introduction

Classically, a *bialgebra* is a vector space carrying the structures of both an associative and unital algebra and a coassociative and counital coalgebra. These are required to be compatible in the sense that the comultiplication and the counit of the coalgebra are algebra homomorphisms; equivalently, the multiplication and the unit of the algebra are coalgebra homomorphisms. A *Hopf algebra* is defined as a bialgebra with an additional property which has several equivalent formulations. The most well-known, perhaps, is the existence of a generalized inverse operation, the so-called *antipode* map. There is an extended literature on Hopf algebras; the classical textbooks are [1, 84, 108].

Historically, Hopf algebras found applications first in *algebraic topology* [28, 58, 60] and in *algebraic group theory* [33, 46, 59, 63, 70]; later also in *Lie algebra theory* [82], *algebraic geometry* [44, 45, 78], *operator algebra* [72, 114] and *combinatorics* [3, 96]. The results in these diverse branches of mathematics also turned out to be fruitful in *physics*; e.g. in condensed matter physics and in quantum field theory [38, 55, 89, 97].

For more on the history and the applications of Hopf algebras we refer to [4, 34]. It should be enough to say here that for half a century Hopf algebras have been successfully applied as symmetry objects in various contexts. In most cases, they enter the picture through their category of modules. In these situations a closed monoidal category M—with possible further structure—is naturally provided, which turns out to be equivalent to the category of modules over a suitable Hopf algebra. Note that for this to happen it is necessary that M comes equipped with a strict closed monoidal functor u to the category of vector spaces (arising as the composite of the desired equivalence with the forgetful functor). The procedure of obtaining the Hopf algebra from the category M and the functor u is known as the *Tannaka–Kreĭn reconstruction* [36, 42, 43, 48, 61, 80, 93, 94, 98–100, 113, 118].

© Springer Nature Switzerland AG 2018
G. Böhm, *Hopf Algebras and Their Generalizations from a Category Theoretical Point of View*, Lecture Notes in Mathematics 2226,
https://doi.org/10.1007/978-3-319-98137-6_1

Although Hopf algebra theory has been a highly successful and popular topic, in various applications some generalizations turned out to be needed. There are many such generalizations which apparently go in different directions. Using the language of Tannaka–Kreĭn duality, this can be understood as the occurrence of more general closed monoidal categories M; potentially ones admitting no strict closed monoidal functor to the category of vector spaces. Clearly, such a category M cannot be interpreted as the category of modules over any Hopf algebra.

In the closest case there is a strict closed monoidal functor to some other braided—or in particular symmetric—closed monoidal category (see e.g. [98, Section I.6.1.3]): say, the category of modules over some commutative ring, the category of graded vector spaces, of simplicial vector spaces, of Hilbert spaces, and so on. A functor of this kind is provided by the forgetful functor from the category of modules over some *Hopf monoid* in the occurring braided monoidal category. A Hopf monoid is no longer an algebra and a coalgebra on a vector space but on a module over a commutative ring, on a graded vector space, on a simplicial vector space or on a Hilbert space, and so on; which means that all structure maps are morphisms in this category. All of these examples can be treated simultaneously using the formalism of braided monoidal categories [62].

Going even further, categories more general than braided monoidal may occur. The categories discussed in [3, Chapter 6] have two different, but compatible monoidal structures. In [3] they were termed 2-monoidal categories; since then (following [107]) they have more often been called *duoidal categories*. This includes braided monoidal categories, where both monoidal structures coincide; in general their compatibility generalizes the braiding. A strict monoidal forgetful functor to one of the monoidal categories underlying the duoidal category is available from module categories over *bimonoids* [3, Section 6.5.1]—whose monoid structure is defined in terms of one monoidal structure and the comonoid structure is defined in terms of the other.

This is still not the end of the story. For instance in subfactor theory [52, 53, 66, 85, 90] and quantum field theory [10, 39, 92] categories occurred from which there is a strict closed monoidal functor to the category of bimodules over some (possibly non-commutative) algebra R. In general categories of bimodules there is no braiding and not even a different second compatible monoidal structure; consequently no bimonoids in them can be considered. Instead, a strict closed monoidal forgetful functor to the category of R-bimodules exists from the category of modules of a *Hopf algebroid* [101, 115] over the base algebra R.

Both from the point of view of the applications [54, 86, 87] and the mathematical treatment, particularly interesting is the case when the base algebra R of a Hopf algebroid T possesses a so-called *separable Frobenius* structure. Then the above strict closed monoidal functor from the category of T-modules to the category of R-bimodules can be composed with the forgetful functor to the category of vector spaces. Although the resulting functor is no longer strict monoidal, thanks to the separable Frobenius structure of R it carries a rich structure [112]. It amounts to both algebra and coalgebra structures on T. However, they no longer constitute a Hopf algebra; e.g. the comultiplication does not preserve the unit. Some weaker axioms of *weak Hopf algebra* [22] hold instead.

In another direction of generalization one deals with a closed monoidal category and a monoidal functor from it to the category of vector spaces. However, the functor is no longer *strict* monoidal (and neither is it assumed to factorize through a strict monoidal one as in the previous paragraph). This situation is realized by the forgetful functor from the category of modules over a *quasi-Hopf algebra* [49]—a variant of Hopf algebra which is not *strictly* coassociative, only up to an invertible cocycle. Although quasi-Hopf algebra has been the subject of intensive investigation and has been successfully applied in mathematical and physical problems [47, 75, 76, 119], it will *not* be mentioned in this volume because it does not fit the philosophy described next.

Although at the level of the explicit forms of the axioms the above generalizations of Hopf algebra look rather different, they share an essential feature explained above: the structure of their category of modules. In all cases—except the not discussed quasi-Hopf algebras of the previous paragraph—it is a closed monoidal category and this structure is strictly preserved by some forgetful functor (the listed examples differ by the target of the forgetful functor). In this book we not only want to demonstrate how the axioms imply this property of the category of modules but—more importantly—also to *derive* the axioms from this requirement.

This aim will be achieved by a fully *category theoretical* treatment. We regard the category of modules over an algebra—and its various generalizations—as the Eilenberg–Moore category of an induced monad. Based on that, the assertion that the *monoidal structure* of a module category is *preserved* by the forgetful functor can be equivalently re-stated as saying that the monoidal structure of the base category is *lifted* to the Eilenberg–Moore category. Then we can apply the well-established theory in [105] of lifting functors and natural transformations to Eilenberg–Moore categories.

The general theory of lifting will be applied first to the particular functors and natural transformations rendering monoidal the base category of a monad t. This results in a bijection between the liftings of the monoidal structure to the Eilenberg–Moore category of t on the one hand; and the opmonoidal structures on t [79, 83] on the other. Recognizing by this the essential role of opmonoidal monads, all structures of

- a *bialgebra* on a vector space
- an *R-bialgebroid* on an $R \otimes R^{op}$-bimodule for an arbitrary algebra R (including, in particular, *weak bialgebra* if R is separable Frobenius)
- a *bimonoid* on an object of a duoidal (so, in particular, braided monoidal) category

will be interpreted as an opmonoidal monad—a.k.a. *bimonad*—structure on a suitable endofunctor.

The next step is a discussion of the preservation; equivalently, lifting of the *closed* structure. This can be done again at the general level of an arbitrary bimonad on a closed monoidal category. Then the closed structure lifts to the Eilenberg–Moore category if and only if a canonical natural transformation is invertible [30, 37].

However, this natural transformation is available for any bimonads on not necessarily closed monoidal categories (although it no longer has a transparent meaning). With its help a *Hopf monad* on any monoidal category can be defined as a bimonad for which it is invertible. A bimonad induced by any of a bialgebra, weak bialgebra, bialgebroid or bimonoid T will be shown to be a Hopf monad if and only if T is a *Hopf algebra, weak Hopf algebra, Hopf algebroid* or bimonoid satisfying a Hopf-like condition, respectively.

By this description of various generalizations of Hopf algebra as instances of the unifying notion of *Hopf monad*, we give a deep explanation of the behavior of their modules; that is, of the closed monoidal structure of the category of modules and its preservation by the forgetful functor.

We intend a pedagogical and self-contained presentation. We begin in Chap. 2 by introducing the necessary theoretical background. Basic notions like *category, functor, natural transformation, adjunction, monad, Eilenberg–Moore algebra* and *lifting* are introduced. All definitions are illustrated by several examples. The choice of the examples is motivated, on one hand, by the wish to help the inexperienced reader to get acquainted with these notions. But they also serve, on the other hand, as preparation for the applications in the later chapters. Although we tried to include in Chap. 2 all the definitions that are relied upon later, it might be too concise for readers without some routine in dealing with them. For more information we recommend the rich supply of introductory books on category theory, for instance the classic textbook [74].

In Chap. 3 *monoidal categories* are introduced and monads on them are studied. The general theory of lifting in Chap. 2 is applied to the particular functors and natural transformations constituting the monoidal structure of a category. This results in a bijection in Theorem 3.19 between the liftings of the monoidal structure of a category A to the Eilenberg–Moore category of a monad t on A; and the opmonoidal structures on t. (A monad with a compatible opmonoidal structure is called succinctly a *bimonad*.) Theorem 3.27 gives a sufficient and necessary condition on an opmonoidal monad t on a *closed* monoidal category A for the closed structure of A to lift to the Eilenberg–Moore category of t. Since this condition is formally meaningful for any bimonad on an arbitrary (not necessarily closed) monoidal category, it is used to define a *Hopf monad* on any monoidal category.

Starting in Chap. 4, certain bimonads and Hopf monads on distinguished categories of particular interest are analyzed. This leads to a unified description of the algebraic structures below.

The base category throughout Chap. 4 is the closed monoidal category **vec** of vector spaces. Those functors **vec** $\to$ **vec** are discussed which are given by taking the tensor product $A \otimes -$ with some vector space A. The bijection between linear maps and natural transformations between such functors results in a bijection between the algebra structures on the vector space A and the monad structures on the functor $A \otimes - : $ **vec** $\to$ **vec**, see Proposition 4.1. A bijection is proven also between the coalgebra structures on the vector space A and the opmonoidal structures on the functor $A \otimes - : $ **vec** $\to$ **vec**, see Proposition 4.3. A combination of these correspondences yields a bijection in Theorem 4.8 between the bialgebras A and the bimonads of the form $A \otimes -$ on vec; which is shown to restrict to a bijection between the Hopf algebras A and the Hopf monads $A \otimes -$.

A similar, but more general, analysis is carried out in Chap. 5 for the closed monoidal category bim(R) of bimodules over an arbitrary (associative and unital but not necessarily commutative) algebra R. Identifying R-bimodules with the left modules of the algebra $R \otimes R^{\mathrm{op}}$, via the $R \otimes R^{\mathrm{op}}$-module tensor product any $R \otimes R^{\mathrm{op}}$-bimodule A induces a functor $A \otimes_{R \otimes R^{\mathrm{op}}} - :$ bim(R) $\to$ bim(R). The bijection between the $R \otimes R^{\mathrm{op}}$-bimodule maps and the natural transformations between such functors results in a bijection between the monad structures on the functor $A \otimes_{R \otimes R^{\mathrm{op}}} - :$ bim(R) $\to$ bim(R) and the algebra homomorphisms $R \otimes R^{\mathrm{op}} \to A$, see Proposition 5.2. A bijection is proven also between the opmonoidal structures on functor $A \otimes_{R \otimes R^{\mathrm{op}}} -$ and the so-called $R|R$-coring structures on A, see Proposition 5.5. A combination of these correspondences yields a bijection in Theorem 5.9 between the bimonads of the form $A \otimes_{R \otimes R^{\mathrm{op}}} -$ on bim(R) and the R-bialgebroids A; which is shown to restrict to a bijection between the Hopf monads $A \otimes_{R \otimes R^{\mathrm{op}}} -$ and the R-Hopf algebroids A.

A *separable Frobenius functor* between monoidal categories is both monoidal and opmonoidal and there are some specific compatibility conditions between these structures. A *separable Frobenius algebra* is both an algebra and a coalgebra and there are some specific compatibility conditions between these structures. In Chap. 6 we first establish in Proposition 6.11 a bijection between the separable Frobenius structures on an algebra R; and separable Frobenius structures on the monoidal forgetful functor bim(R) $\to$ vec. This is used to improve the results of Chap. 5 in the particular case of a separable Frobenius base algebra. For any algebra A, Theorem 6.19 asserts bijections between the bialgebroid structures on A over some separable-Frobenius algebra; the weak bialgebra structures on A; and those monoidal structures on the category of A-modules for which the forgetful functor to vec is a separable Frobenius functor. Seen as a bialgebroid, A is a Hopf algebroid if and only if, as a weak bialgebra, it is a weak Hopf algebra.

In Chap. 7 we work with unspecified monoidal and duoidal categories. In an arbitrary monoidal category C with monoidal product $\otimes$, not every natural transformation $A \otimes - \to A' \otimes -$ between functors C $\to$ C is induced by a morphism $A \to A'$. Hence no bijection can be expected between the monoids A and the monads $A \otimes -$. Their precise relation is described in Proposition 7.6. If C is not merely a monoidal but a duoidal category with monoidal products $\diamond$ and $\blacklozenge$, then we may ask about the opmonoidal functors $A \blacklozenge - :$ (C, $\diamond$) $\to$ (C, $\diamond$). In Proposition 7.13 a bijection is verified between the comonoids A in (C, $\diamond$) and some distinguished opmonoidal functors $A \blacklozenge - :$ (C, $\diamond$) $\to$ (C, $\diamond$). Combining these results, in Theorem 7.18 a bijection is proven between the bimonoids A in the duoidal category (C, $\diamond$, $\blacklozenge$) and the induced bimonads $A \blacklozenge -$ on (C, $\diamond$). Those bimonoids are identified whose induced bimonad is a Hopf monad. Bimonoids in duoidal categories include several interesting examples; such as the bimonoids in braided—so in particular symmetric—monoidal categories (which induce a Hopf monad if and only if they admit an antipode; i.e. they are a Hopf monoid), small categories (which induce a Hopf monad if and only if they are a groupoid), bialgebroids over *commutative* base algebras (which induce a Hopf monad if and only if they are a Hopf algebroid), and weak bialgebras (which induce a Hopf monad if and only if they are a weak Hopf algebra). General bialgebroids, however, are not known to fit the duoidal category picture.

Some details along the way are left to the readers as exercises, with the hope that working them out will lead to a deeper understanding. However, since some of our proofs rely on them, for completeness the solutions to the exercises are included in Chap. 8.

Chapter 2
Lifting to Eilenberg–Moore Categories

This chapter aims to present the necessary background in category theory. The structures occurring in the later sections are introduced and their key properties are discussed. All of the definitions are illustrated by collections of examples, chosen for their relevance to the applications in the later sections.

We begin with some basic notions like *category*, *functor*, *natural transformation* and operations with them. Then we turn to *adjunctions* and *monads*. The *Eilenberg–Moore category* of a monad is defined together with the key concept of *lifting* of functors, natural transformations and adjunctions to Eilenberg–Moore categories of monads.

Our presentation is necessarily concise; its main aim is to fix the notation and terminology. For more details, we refer to [74].

Definition 2.1 ([74, Section I.8]). A *category* A consists of

- a class of *objects* $X, Y, \ldots$
- for each pair of objects X, Y a collection $A(X, Y)$ of *morphisms* $X \to Y$ (where X is said to be the *source* and Y is said to be the *target* of a morphism $X \to Y$)
- for each object X a map $\mathbf{1}$ from the singleton set $\mathbb{1}$ to $A(X, X)$ (whose image is termed the *identity morphism* $X \to X$)
- for each triple of objects X, Y, Z a map $\circ$ from the Cartesian product set $A(Y, Z) \times A(X, Y)$ to $A(X, Z)$ (termed the *composition*)

such that for all objects X, Y, Z, V the following diagrams commute.

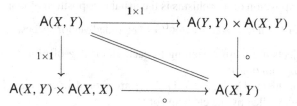

© Springer Nature Switzerland AG 2018
G. Böhm, *Hopf Algebras and Their Generalizations from a Category Theoretical Point of View*, Lecture Notes in Mathematics 2226,
https://doi.org/10.1007/978-3-319-98137-6_2

$$A(Z, V) \times A(Y, Z) \times A(X, Y) \xrightarrow{\circ \times 1} A(Y, V) \times A(X, Y)$$

$$\downarrow{\scriptstyle 1 \times \circ} \qquad\qquad\qquad\qquad\qquad \downarrow{\scriptstyle \circ}$$

$$A(Z, V) \times A(X, Z) \xrightarrow{ \circ } A(X, V)$$

A category is said to be *small* if its morphisms constitute a set (whence also the subclass of identity morphisms—which is in bijection with the objects—is a set too).

In order to make a clear notational difference from the map assigning identity morphisms to the objects, denoted by a boldface character **1**, the regular character 1 is used to denote identity maps.

Example 2.2.

1. Any set X can be seen as a category in which the objects are the elements of X and in which there are only identity morphisms. It is called the *discrete category* on the set X and it is denoted by $\mathsf{disc}(X)$, or simply by X if there is no risk of confusion.

    In particular, the *singleton category* $\mathbb{1}$ consists of a single object and its identity morphism.

2. Any set X can be seen as the set of objects in another category $\mathsf{ind}(X)$, the so-called *indiscrete category* in which there is precisely one morphism $x \to y$ for any objects x and y.

3. In the category set of sets,

    - the objects are sets $X, Y, \ldots$
    - the morphisms from X to Y are the maps $X \to Y$
    - the identity morphism $X \to X$ is the identity map
    - the composition of morphisms is the usual composition of maps.

4. In the category vec of vector spaces,

    - the objects are vector spaces (over a given field) $X, Y, \ldots$
    - the morphisms from X to Y are the linear maps $X \to Y$
    - the identity morphism $X \to X$ is the identity map
    - the composition of morphisms is the usual composition of maps.

5. For any set G, in the category vec^G of G-graded vector spaces,

    - the objects are maps assigning to each element g of G a vector space X_g (over a given field)
    - the morphisms $\{X_g\}_{g \in G} \to \{Y_g\}_{g \in G}$ are sets of linear maps $\{f_g : X_g \to Y_g\}_{g \in G}$ labelled by the elements of G
    - the identity morphism $\{X_g\}_{g \in G} \to \{X_g\}_{g \in G}$ consists of identity maps $\{1 : X_g \to X_g\}_{g \in G}$

- composition is defined componentwise;

$$\{h_g : Y_g \to Z_g\}_{g \in G} \circ \{f_g : X_g \to Y_g\}_{g \in G} := \{h_g \circ f_g : X_g \to Z_g\}_{g \in G}.$$

6. Consider an *algebra*; that is, a vector space A over a given field k, equipped with linear maps $m : A \otimes A \to A$—called the *multiplication*—and $i : k \to A$—called the *unit*—such that m is associative in the sense that $m \circ (m \otimes 1) = m \circ (1 \otimes m)$ and unital in the sense that $m \circ (i \otimes 1) = 1 = m \circ (1 \otimes i)$. On elements of A multiplication is often denoted by juxtaposition and the unit element—that is, the image of the multiplicative unit of k under the map i—is denoted by 1.

   In the category $\mathsf{mod}(A)$ of left A-modules,

   - the objects are left A-*modules*; that is, vector spaces X equipped with a linear map $x : A \otimes X \to X$—called the *action*—which is associative in the sense that $x \circ (1 \otimes x) = x \circ (m \otimes 1)$, and unital i.e. $x \circ (i \otimes 1) = 1$
   - the morphisms are the A-module homomorphisms; that is, the linear maps $f : X \to X'$ such that $f \circ x = x' \circ (1 \otimes f)$
   - the identity morphism $X \to X$ is the identity map
   - the composition of morphisms is the usual composition of maps.

7. In the category hil of Hilbert spaces,

   - the objects are the Hilbert spaces over the field of complex numbers
   - the morphisms are the continuous linear maps (not required to preserve the inner product)
   - the identity morphisms are the identity maps
   - composition is the usual composition of maps.

8. In the category alg of algebras over a given field,

   - the objects are the associative and unital algebras
   - the morphisms are the algebra homomorphisms; that is, multiplicative and unital linear maps
   - the identity morphisms are the identity maps
   - the composition of morphisms is the usual composition of maps.

   There is an analogous category coalg of coalgebras and coalgebra maps over a given field—see more about them in Paragraph 4.2 below.

9. To any category A we can associate its *opposite* A^{op} in which

   - the objects are the same as the objects of A
   - the morphisms $X \nrightarrow Y$ are the morphisms $Y \to X$ in A
   - the identity morphisms are the same as in A
   - the composition of morphisms is the opposite of that in A; that is, the composite of $f : X \nrightarrow Y$ and $g : Y \nrightarrow Z$ is $Z \xrightarrow{g} Y \xrightarrow{f} X : X \nrightarrow Z$.

10. The *Cartesian product* of any categories A and B is the category A × B in
    which

    - the objects are pairs of an object X of A and an object Y of B
    - the morphisms $(X, Y) \to (X', Y')$ are pairs of a morphism $X \to X'$ in A
      and a morphism $Y \to Y'$ in B
    - the identity morphisms are pairs of identity morphisms $(1 : X \to X, 1 : Y \to Y)$
    - the composition is defined memberwise: $(f', g') \circ (f, g) = (f' \circ f, g' \circ g)$.

Definition 2.3 ([74, Section I.5]). A morphism f in a category is said to be a
monomorphism if $f \circ h = f \circ h'$ implies $h = h'$ for any morphisms h and h' whose
target is equal to the source of f.

A *split monomorphism* is a morphism $f : X \to Y$ which admits a left inverse;
that is, a morphism $g : Y \to X$ such that $g \circ f$ is the identity morphism of X.
Clearly, a split monomorphism is a monomorphism.

Dually, f is said to be an *epimorphism* if it is a monomorphism in the opposite
category (cf. Example 2.2 9); that is, $h \circ f = h' \circ f$ implies $h = h'$ for any morphisms
h and h' whose source is equal to the target of f.

A *split epimorphism* is a morphism $f : X \to Y$ which admits a right inverse;
that is, a morphism $g : Y \to X$ such that $f \circ g$ is the identity morphism of Y (thus
a split epimorphism is an epimorphism).

An *isomorphism* is an invertible morphism; that is, a morphism $f : X \to Y$ for
which there exists a morphism—the *inverse*—$g : Y \to X$ such that both composites
$f \circ g$ and $g \circ f$ are identity morphisms (so that f and g are split monomorphisms
and split epimorphisms as well). In this case the objects X and Y are said to be
isomorphic.

Definition 2.4 ([74, Section I.8]). A *functor* f from a category A to a category B
consists of

- a map associating to each object X of A an object fX of B
- for each pair of objects X, Y a map $f : A(X, Y) \to B(fX, fY)$

such that the following diagrams commute for any objects X, Y, Z of A.

$$
\begin{array}{ccc}
\mathbb{1} & \xrightarrow{1} & A(X, X) \\
\| & & \downarrow f \\
\mathbb{1} & \xrightarrow{1} & B(fX, fX)
\end{array}
\qquad
\begin{array}{ccc}
A(Y, Z) \times A(X, Y) & \xrightarrow{\circ} & A(X, Z) \\
f \times f \downarrow & & \downarrow f \\
B(fY, fZ) \times B(fX, fY) & \xrightarrow{\circ} & B(fX, fZ)
\end{array}
$$

A functor $f : A \to B$ is said to be *faithful* if the induced map $A(X, Y) \to B(fX, fY)$ is injective for all objects X, Y.

The (evident) composition of functors $t : A \to B$ and $s : B \to C$ will be denoted by juxtaposition $st : A \to C$. For repeated (n fold) application of the same functor $t : A \to A$ the power notation t^n is also used. The identity functor $A \to A$ will be denoted by 1.

Example 2.5.

1. Regarding any vector space as a plain set, and regarding linear maps as plain maps of the underlying sets we obtain a *forgetful functor* $u :$ vec $\to$ set.
2. In the opposite direction, for any given field k we can take the vector space kX spanned by the elements of a fixed set X. Since any map $X \to Y$ extends to a linear map $kX \to kY$, this yields a 'linearization' functor $k :$ set $\to$ vec.
3. If A is an algebra over a field k then every A-module is in particular a vector space over k and A-module maps are in particular k-linear. This again yields a forgetful functor $\mathsf{mod}(A) \to$ vec.
4. Let us take next algebras A and B over a field k and a B-A bimodule W; that is, a left B-action $l : B \otimes W \to W$ and a right A-action $r : W \otimes A \to W$ which commute in the sense that $l \circ (1 \otimes r) = r \circ (l \otimes 1)$. For any left A-module V we can take the A-module tensor product $W \otimes_A V$, which is the quotient of the vector space $W \otimes V$ by the subspace

$$\{w \cdot a \otimes v - w \otimes a \cdot v \mid a \in A,\ v \in V,\ w \in W\}.$$

   Via the B-action on W, $W \otimes_A V$ is a left B-module, and for any left A-module map $h : V \to V'$ there is a left B-module map $1 \otimes_A h : W \otimes_A V \to W \otimes_A V'$. This defines a functor $W \otimes_A - : \mathsf{mod}(A) \to \mathsf{mod}(B)$. In particular, for any left B-module W there is a functor $W \otimes - :$ vec $\to \mathsf{mod}(B)$ and thus for any vector space W there is a functor $W \otimes - :$ vec $\to$ vec.
5. In the opposite direction, in terms of the same data A, B and W, for any left B-module Z we can regard $\mathsf{mod}(B)(W, Z)$—that is, the set of B-module maps $W \to Z$—as a vector space with the pointwise linear structure $(q + q')(w) := q(w) + q'(w)$ and $(\lambda q)(w) := \lambda q(w)$ for all B-module maps $q, q' : W \to Z$, $w \in W$ and $\lambda \in k$. This vector space $\mathsf{mod}(B)(W, Z)$ is a left A-module with action $(a \cdot q)(w) := q(w \cdot a)$ for a B-module map $q : W \to Z$, $a \in A$ and $w \in W$. Post-composition with any B-module map $l : Z \to Z'$ yields an A-module map $l \circ - : \mathsf{mod}(B)(W, Z) \to \mathsf{mod}(B)(W, Z')$ defining the functor $\mathsf{mod}(B)(W, -) : \mathsf{mod}(B) \to \mathsf{mod}(A)$. In particular, for any left B-module W there is a functor $\mathsf{mod}(B)(W, -) : \mathsf{mod}(B) \to$ vec and thus for any vector space W there is a functor $\mathsf{vec}(W, -) :$ vec $\to$ vec.
6. For any category A, functors from the singleton category $\mathbb{1}$ in Example 2.2 1 to A are in a bijective correspondence with the objects of A (i.e. with the images of the single object of $\mathbb{1}$).
7. Any functor $f : A \to B$ can be seen as a functor $f^{\mathsf{op}} : A^{\mathsf{op}} \to B^{\mathsf{op}}$ between the opposite categories of Example 2.2 9.

Definition 2.6 ([74, Section I.4]).   Consider functors f and g of equal source A and equal target B (we say that f and g are *parallel* functors). A *natural transformation* $\varphi : f \to g$ consists of

- for each object X of A a morphism $\varphi_X : fX \to gX$ in B

such that the following diagram commutes for any morphism $h : X \to Y$ in A.

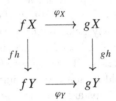

This diagram is said to express the *naturality* of φ with respect to h.

Example 2.7.

1. Any algebra B can be seen as a B-bimodule with actions provided by the multiplication. Consequently, as in part 5 of Example 2.5, there is a functor $\mathsf{mod}(B)(B, -) : \mathsf{mod}(B) \to \mathsf{mod}(B)$. For any left B-module Z, the B-module map from $\mathsf{mod}(B)(B, Z)$ to Z, provided by the evaluation of a B-module map $q : B \to Z$ on the unit element of the algebra B, and its inverse $Z \to \mathsf{mod}(B)(B, Z)$ sending z to the map $b \mapsto b \cdot z$ define natural transformations between $\mathsf{mod}(B)(B, -) : \mathsf{mod}(B) \to \mathsf{mod}(B)$ and the identity functor.
2. Composing the functors in parts 1 and 2 of Example 2.5 we obtain a functor sending a set X to the set of elements in the vector space kX. The evident inclusion maps $X \to kX$ yield the components of a natural transformation from the identity functor $\mathsf{set} \to \mathsf{set}$ to this composite functor.
3. In the situation of part 4 of Example 2.5, take a B-A bimodule map $p : W \to W'$—that is, a linear map which is compatible with both actions. Then for any left A-module V, the B-module maps $p \otimes_A V : W \otimes_A V \to W' \otimes_A V$ yield the components of a natural transformation $W \otimes_A - \to W' \otimes_A -$.
4. In the same setting as in the previous item 3, pre-composition with p defines a left A-module map $- \circ p : \mathsf{mod}(B)(W', Z) \to \mathsf{mod}(B)(W, Z)$ for any left B-module Z. These maps can be seen as the components of a natural transformation $\mathsf{mod}(B)(W', -) \to \mathsf{mod}(B)(W, -)$.
5. Composing the functors in parts 4 and 5 of Example 2.5 we obtain a functor sending a left A-module V to the A-module $\mathsf{mod}(B)(W, W \otimes_A V)$. The maps $V \to \mathsf{mod}(B)(W, W \otimes_A V)$ sending v to the map $w \mapsto w \otimes_A v$ yield the components of a natural transformation from the identity functor $\mathsf{mod}(A) \to \mathsf{mod}(A)$ to $\mathsf{mod}(B)(W, W \otimes_A -)$.
6. Composing the functors in parts 4 and 5 of Example 2.5 in the opposite order, we obtain a functor sending a left B-module Z to the B-module $W \otimes_A \mathsf{mod}(B)(W, Z)$. The evaluation maps $W \otimes_A \mathsf{mod}(B)(W, Z) \to Z$ sending $w \otimes_A q$ to $q(w)$ yield the components of a natural transformation from $W \otimes_A \mathsf{mod}(B)(W, -)$ to the identity functor $\mathsf{mod}(B) \to \mathsf{mod}(B)$.

2.9. Operations with Natural Transformations. [74, Sections II.4 and II.5] For natural transformations $\varphi : f \to f'$ and $\varphi' : f' \to f''$ between parallel functors $\mathsf{A} \to \mathsf{B}$, their *composite* $\varphi' \circ \varphi : f \to f''$ has the following component at any object X of A

$$fX \xrightarrow{\ \varphi_X\ } f'X \xrightarrow{\ \varphi'_X\ } f''X.$$

The identity for this composition is the *identity natural transformation* $1 : f \to f$ with the component at any object X of A the identity morphism

$$fX \xrightarrow{\ 1\ } fX.$$

A natural transformation which has an inverse for this composition is termed a *natural isomorphism*. This is equivalent to saying that each component has an inverse.

For parallel functors $f, f' : \mathsf{A} \to \mathsf{B}$ which are composable with the parallel functors $g, g' : \mathsf{B} \to \mathsf{C}$, and for natural transformations $\varphi : f \to f'$ and $\gamma : g \to g'$, there is a natural transformation $\gamma\varphi : gf \to g'f'$—known as the *Godement product* of φ and γ—with component at any object X of A occurring in the equal paths around the following diagram

$$
\begin{array}{ccc}
gfX & \xrightarrow{\ \gamma_{fX}\ } & g'fX \\
{\scriptstyle g\varphi_X}\downarrow & & \downarrow{\scriptstyle g'\varphi_X} \\
gf'X & \xrightarrow[\ \gamma_{f'X}\ ]{} & g'f'X
\end{array}
$$

Example 2.11.

1. Part 1 of Example 2.7 describes a natural isomorphism between the functor $\mathsf{mod}(B)(B, -)$ and the identity functor $\mathsf{mod}(B) \to \mathsf{mod}(B)$.
2. Composing both naturally isomorphic functors in item 1 above with the forgetful functor $u : \mathsf{mod}(B) \to \mathsf{vec}$ from part 3 of Example 2.5, we get a natural isomorphism between $\mathsf{mod}(B)(B, -) : \mathsf{mod}(B) \to \mathsf{vec}$ and u.

Definition 2.12 ([74, Section IV.4]).   An *equivalence* between some categories A and B consists of

- functors $f : \mathsf{A} \to \mathsf{B}$ and $g : \mathsf{B} \to \mathsf{A}$
- natural isomorphisms $fg \to 1$ and $gf \to 1$.

It is proven in [74, Section IV.4] that a functor $f : \mathsf{A} \to \mathsf{B}$ takes part in an equivalence if and only if it is

(a) *essentially surjective* on the objects; which means that any object Y of B is isomorphic to an object of the form fX for a suitable object X of A and
(b) bijective on the hom sets $\mathsf{A}(X, X') \to \mathsf{B}(fX, fX')$ for all objects X, X' of A.

Example 2.13.

1. If for some functors $f : \mathsf{A} \to \mathsf{B}$ and $g : \mathsf{B} \to \mathsf{A}$ both composites fg and gf are identity functors, then f and g are said to be mutually inverse *isomorphisms*. Then they clearly constitute an equivalence with the identity natural transformations $fg \to 1$ and $gf \to 1$.
2. Any element x of an arbitrary set X determines a functor—also denoted by x— from the singleton category $\mathbb{1}$ of Example 2.2 1 to the indiscrete category $\mathsf{ind}(X)$ of Example 2.2 2 as described in Example 2.5 6. Together with the unique functor $t : \mathsf{ind}(X) \to \mathbb{1}$ they constitute an equivalence. Indeed, tx is the identity functor (we have the identity natural isomorphism $tx \to 1$); and the component of a natural isomorphism $xt \to 1$ at any object y is the unique morphism $x \to y$.
3. Consider a strict *Morita context* with algebras A and B, A-B bimodule M and B-A bimodule N, and bimodule isomorphisms $\varphi : M \otimes_B N \to A$ and $\psi : N \otimes_A M \to B$. Then the functors $N \otimes_A \otimes - : \mathsf{mod}(A) \to \mathsf{mod}(B)$ and $M \otimes_B \otimes - : \mathsf{mod}(B) \to \mathsf{mod}(A)$ of Example 2.5 4, together with the natural isomorphisms

$$M \otimes_B N \otimes_A - \xrightarrow{\ \varphi \otimes_A - \ } A \otimes_A - \xrightarrow{\ \cong\ } 1 \quad \text{and} \quad N \otimes_A M \otimes_B - \xrightarrow{\ \psi \otimes_B - \ } B \otimes_B - \xrightarrow{\ \cong\ } 1,$$

constitute an equivalence.

Equivalence functors preserve many relevant features of a category (but not the number of objects or morphisms; see Example 2.13 2). So as long as we are interested in properties and structures preserved by equivalence functors, we can choose—e.g. by convenience—which one of some equivalent categories to work with.

Definition 2.14 ([74, Section IV.1]). An *adjunction* consists of

- functors $r : A \to B$ and $l : B \to A$
- for each object X of B and Y of A an isomorphism (i.e. bijection) $A(lX, Y) \cong B(X, rY)$ which are natural in both objects X and Y; that is, for any morphisms $p : X' \to X$ in B and $q : Y \to Y'$ in A the following diagram commutes.

$$
\begin{array}{ccc}
A(lX, Y) & \xrightarrow{\ \cong\ } & B(X, rY) \\
{\scriptstyle q\circ-\circ lp}\Big\downarrow & & \Big\downarrow{\scriptstyle rq\circ-\circ p} \\
A(lX', Y') & \xrightarrow[\ \cong\ ]{} & B(X', rY')
\end{array}
$$

$$(2.1)$$

An adjunction is denoted by $l \dashv r : A \to B$ (without explicitly referring to the natural isomorphism part). We say that l is the *left adjoint* (of r) and r is the *right adjoint* (of l).

Proposition 2.15 ([74, IV.1 Theorem 2]). *An adjunction can equivalently be described by the following data*

- *functors $r : A \to B$ and $l : B \to A$*
- *natural transformations $\eta : 1 \to rl$ and $\varepsilon : lr \to 1$*

such that the following diagrams commute; that is, the so-called triangle identities *hold.*

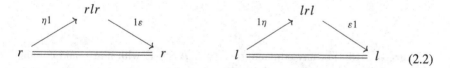

$$(2.2)$$

The natural transformation η is called the unit *and ε is called the* counit *of the adjunction.*

Proof. Suppose that natural isomorphisms $\xi_{X,Y} : A(lX, Y) \to B(X, rY)$ as in Definition 2.14 are given. The component of η at any object X of B is then constructed as $\xi_{X,lX}(1)$ and the component of ε at any object Y of A is constructed as $\xi^{-1}_{rY,Y}(1)$. The component of the upper path of the first diagram in (2.2) at any object Y of A is equal to $r\xi^{-1}_{rY,Y}(1) \circ \xi_{rY,lrY}(1)$. By the naturality condition (2.1) this is equal to $\xi_{rY,Y}(\xi^{-1}_{rY,Y}(1))$ thus to the identity morphism $rY \to rY$. The second triangle identity of (2.2) is checked symmetrically. The naturality of η follows by applying (2.1) twice: for any morphism $t : X \to X'$ in B,

$$rlt \circ \xi_{X,lX}(1) = \xi_{X,lX'}(lt) = \xi_{X',lX'}(1) \circ t.$$

The naturality of ε is checked analogously.

Conversely, suppose that natural transformations η and ε as in the claim are given. The natural isomorphism in Definition 2.14 is constructed as

$$A(lX, Y) \to B(X, rY), \qquad h \mapsto rh \circ \eta_X.$$

Using the naturality of η and ε together with the triangle identities (2.2) we see that it has the inverse

$$B(X, rY) \to A(lX, Y), \qquad h' \mapsto \varepsilon_Y \circ lh'.$$

The commutativity of (2.1) follows by the naturality of η and the functoriality of r: for any morphism $h : lX \to Y$ in A,

$$rq \circ rh \circ \eta_X \circ p = rq \circ rh \circ rlp \circ \eta_{X'} = r(q \circ h \circ lp) \circ \eta_{X'}.$$

It is immediate to see that the above constructions are mutual inverses.   □

Example 2.16. The natural transformations in parts 5 and 6 of Example 2.7 are the unit and the counit, respectively, of an adjunction $W \otimes_A - \dashv \mathsf{mod}(B)(W, -) : \mathsf{mod}(B) \to \mathsf{mod}(A)$ induced by the B-A bimodule W.

Exercise 2.17. Show that whenever a functor possesses a (left or right) adjoint, it is unique up to a natural isomorphism; and this isomorphism is compatible with the unit and the counit of the adjunction.

Exercise 2.18. For any adjunction $l \dashv r : A \to B$ with unit η and counit ε, verify bijective correspondences between natural transformations of the following kinds.

(1) Between natural transformations $fr \to g$ and $f \to gl$, for any functors $f : B \to C$, $g : A \to C$ and any category C.
(2) Between natural transformations $f \to rg$ and $lf \to g$, for any functors $f : C \to B$, $g : C \to A$ and any category C.

Exercise 2.19. Prove that if the functors $f : A \to B$ and $g : B \to A$ take part in an equivalence (with the natural isomorphisms $\varphi : fg \to 1$ and $\psi : gf \to 1$) then f is both the left adjoint and the right adjoint of g; moreover, the unit and the counit of both adjunctions are isomorphisms.

Definition 2.20 ([74, Section VI.1]).   A *monad* on a category A consists of

- a functor $t : A \to A$
- a natural transformation η from the identity functor 1 to t (known as the *unit* of the monad)
- a natural transformation μ from the twofold iterate t^2 to t (known as the multiplication of the monad)

such that the following diagrams commute.

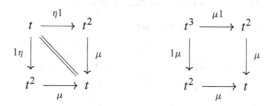

Example 2.21.

1. Every identity functor can be made a monad with the identity natural transformation as the multiplication and the unit.

2. Consider an adjunction $l \dashv r : A \to B$ (with unit $\eta : 1 \to rl$ and counit $\varepsilon : lr \to 1$) and a monad t on A (with unit $\iota : 1 \to t$ and multiplication $\mu : t^2 \to t$). Then there is a monad on B with functor part rtl; unit and multiplication

$$1 \xrightarrow{\ \eta\ } rl \xrightarrow{\ 1\iota 1\ } rtl \qquad\qquad rtlrtl \xrightarrow{\ 1 1\varepsilon 1 1\ } rt^2 l \xrightarrow{\ 1\mu 1\ } rtl.$$

Indeed, associativity of the multiplication is immediate by the naturality of ε and the associativity of μ; and unitality follows by the unitality of μ and the triangle conditions.

In particular, taking t to be the identity monad in item 1, we get a monad structure on the functor rl. This monad is said to be *induced by the adjunction* $l \dashv r$.

3. Consider an algebra A over a field k with unit map $i : k \to A$ and multiplication $m : A \otimes A \to A$. By Example 2.16 and part 2 of Example 2.11 there is an adjunction $A \otimes - \dashv u : \mathsf{mod}(A) \to \mathsf{vec}$ whose induced monad—as in item 2 above—lives on the functor $A \otimes - : \mathsf{vec} \to \mathsf{vec}$ (seen in part 4 of Example 2.5). The unit of this monad is the natural transformation from the identity functor to $A \otimes -$ with components $i \otimes 1 : V \to A \otimes V$, for any vector space V. The multiplication is the natural transformation $A \otimes A \otimes - \to A \otimes -$ with components $m \otimes 1 : A \otimes A \otimes V \to A \otimes V$ for any vector space V.

4. Consider a monad (t, η, μ) and a natural isomorphism $\varphi : t \to s$. There is a monad s with the multiplication and unit

$$ss \xrightarrow{\ \varphi^{-1}\varphi^{-1}\ } tt \xrightarrow{\ \mu\ } t \xrightarrow{\ \varphi\ } s \qquad\qquad 1 \xrightarrow{\ \eta\ } t \xrightarrow{\ \varphi\ } s.$$

A natural question arises at this point whether any monad is induced by an adjunction (in the way discussed in the last paragraph of part 2 of Example 2.21). This indeed turns out to be the case, as we shall discuss next.

Definition 2.22 ([74, Section VI.2]). Consider a monad t on a category A with unit η and multiplication μ. An *Eilenberg–Moore algebra (or module)* over this monad consists of

- an object V of A
- a morphism $v : tV \to V$ in A (the so-called *action*)

such that the following diagrams commute

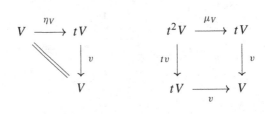

Together with the morphisms $h : V \to V'$ such that $v' \circ th = h \circ v$, the Eilenberg–Moore algebras of t constitute the so-called *Eilenberg–Moore category*, denoted as A^t.

Example 2.23.

1. Regard an identity functor $1 : \mathsf{A} \to \mathsf{A}$ as a monad, in the way described in part 1 of Example 2.21. Since its Eilenberg–Moore algebras must be unital, they must be of the form $(X, \mathbf{1})$ for any object X of A. Consequently, the Eilenberg–Moore category is A itself.
2. Consider the monad in part 3 of Example 2.21, induced by an algebra A. Its Eilenberg–Moore category is the category $\mathrm{mod}(A)$ of left A-modules.

2.24. The Eilenberg–Moore Adjunction [74, VI.2 Theorem 1] For any monad (t, η, μ) on A, there is a *forgetful functor* u^t from the Eilenberg–Moore category A^t of Definition 2.22 to the base category A. It sends an Eilenberg–Moore algebra (V, v) to the constituent object V and it acts on the morphisms as the identity map.

The forgetful functor u^t has a left adjoint f^t sending an object X of A to the Eilenberg–Moore algebra (tX, μ_X) and sending a morphism h to th. The unit of the adjunction should be a natural transformation $1 \to u^t f^t = t$; its component at any object X of A is provided by η_X in terms of the unit η of the monad t. The counit of the adjunction should be a natural transformation $f^t u^t \to 1$; its component at any object (V, v) of A^t is $v : (tV, \mu_V) \to (V, v)$.

The monad induced by the above adjunction $f^t \dashv u^t$—in the sense of Example 2.21 2—is equal to (t, η, μ).

The Eilenberg–Moore adjunction of Paragraph 2.24 is not the only adjunction inducing a given monad, as illustrated by the following Exercise.

Exercise 2.25. For any monad (t, η, μ) on some category A consider the so-called *Kleisli category* A_t; whose objects are the same objects of A and whose morphisms $X \nrightarrow Y$ are the morphisms $X \to tY$ in A. Such morphisms are composed by the rule

$$(\; Y \overset{g}{\nrightarrow} Z \;) \circ (\; X \overset{f}{\nrightarrow} Y \;) := (X \overset{f}{\longrightarrow} tY \overset{tg}{\longrightarrow} t^2Z \overset{\mu z}{\longrightarrow} tZ \;)$$

and the identity morphism $\; X \nrightarrow X \;$ is $\eta_X : X \to tX$.

(1) Show that the functor

$$f_t : \mathsf{A} \to \mathsf{A}_t, \qquad (\; X \overset{h}{\longrightarrow} Y \;) \mapsto (\; X \overset{h}{\longrightarrow} Y \overset{\eta_Y}{\longrightarrow} tY \;)$$

is the left adjoint of

$$u_t : \mathsf{A}_t \to \mathsf{A}, \qquad (\; X \overset{g}{\nrightarrow} Y \;) \mapsto (\; tX \overset{tg}{\longrightarrow} t^2Y \overset{\mu_Y}{\longrightarrow} tY \;).$$

(2) Verify that the adjunction of part (1) induces—in the sense of Example 2.21 2—the given monad (t, η, μ).

Although not unique as just seen, the Eilenberg–Moore adjunction of Paragraph 2.24 is distinguished—in the sense of the following Exercise—among the adjunctions inducing the same monad.

Exercise 2.26. Prove that for any adjunction $l \dashv r : \mathsf{B} \to \mathsf{A}$ inducing in the sense of Example 2.21 2 a given monad (t, η, μ) on A, there is a unique functor k from B to the Eilenberg–Moore category A^t of Definition 2.22 for which the following diagram (using the notation of Paragraph 2.24) commutes.

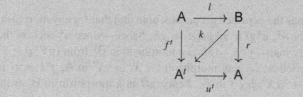

$$\text{(2.3)}$$

Theorem and Definition 2.27 ([105]).   For any monads t on a category A and s on a category B, and any functor $g : \mathsf{A} \to \mathsf{B}$, there is a bijective correspondence between the following data.

(i) Functors $g^\gamma : \mathsf{A}^t \to \mathsf{B}^s$ rendering commutative the following diagram (in which the notation of Paragraph 2.24 is used).

$$
\begin{array}{ccc}
\mathsf{A}^t & \xrightarrow{\ g^\gamma\ } & \mathsf{B}^s \\
{\scriptstyle u^t}\downarrow & & \downarrow{\scriptstyle u^s} \\
\mathsf{A} & \xrightarrow{\ \ g\ \ } & \mathsf{B}
\end{array}
$$

(ii) Natural transformations $\gamma : sg \to gt$ which are compatible with the units η and multiplications μ of both monads, in the sense that the following diagrams commute.

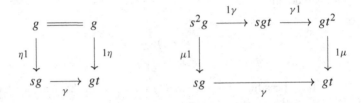

$$
\begin{array}{ccc}
g & =\!=\!= & g \\
{\scriptstyle \eta 1}\downarrow & & \downarrow{\scriptstyle 1\eta} \\
sg & \xrightarrow{\ \gamma\ } & gt
\end{array}
\qquad\qquad
\begin{array}{ccccc}
s^2 g & \xrightarrow{\ 1\gamma\ } & sgt & \xrightarrow{\ \gamma 1\ } & gt^2 \\
{\scriptstyle \mu 1}\downarrow & & & & \downarrow{\scriptstyle 1\mu} \\
sg & & \xrightarrow{\qquad\quad \gamma \qquad\quad} & & gt
\end{array}
$$

The functor g^γ is termed a *lifting* of g (along the *monad morphism* γ.)

Proof. Assume first that the functor g^γ in part (i) is given. It takes any Eilenberg–Moore t-algebra (V, v) to some s-algebra. From the functor equality $u^s g^\gamma = g u^t$ we know that the object part of $g^\gamma (V, v)$ must be $u^s g^\gamma (V, v) = g u^t (V, v) = gV$; which carries then some s-action $\varrho_{(V,v)} : sgV \to gV$ in terms of which $g^\gamma (V, v) = (gV, \varrho_{(V,v)})$. In particular, there is an s-action $\varrho_{(tX,\mu_X)} : sgtX \to gtX$ for any object X of A such that $g^\gamma f^t X = (gtx, \varrho_{(tX,\mu_X)})$.

We construct the components of the desired natural transformation γ as

$$
\gamma_X := \quad sgX \xrightarrow{\ sg\eta_X\ } sgtX \xrightarrow{\ \varrho_{(tX,\mu_X)}\ } gtX
$$

for any object X of A.

In order to see that it has the expected properties, note first that for any morphism $h : (V, v) \to (V', v')$ in A^t, $u^s g^\gamma h = g u^t h = gh$, hence—since u^s acts on the morphisms as the identity map—$g^\gamma h = gh$ is a morphism in B^s from $(gV, \varrho_{(V,v)})$ to $(gV', \varrho_{(V',v')})$. In particular, for any morphism $l : X \to X'$ in A, $f^t l = tl$ is a morphism $(tX, \mu_X) \to (tX', \mu_{X'})$ in A^t, hence gtl is a morphism in B^s from $(gtX, \varrho_{(tX,\mu_X)})$ to $(gtX', \varrho_{(tX',\mu_{X'})})$. Using this, and the naturality of the unit η of the monad t, we see that γ is natural; that is, the following diagram commutes for any morphism $l : X \to X'$ in **A**.

$$sgX \xrightarrow{\ sg\eta_X\ } sgtX \xrightarrow{\ \varrho_{(tX,\mu_X)}\ } gtX$$

$$\Big\downarrow sgl \qquad\qquad \Big\downarrow sgtl \qquad\qquad \Big\downarrow gtl$$

$$sgX' \xrightarrow{\ sg\eta_{X'}\ } sgtX' \xrightarrow{\ \varrho_{(tX',\mu_{X'})}\ } gtX'$$

The compatibility of γ with the units of both monads t and s; that is, commutativity of the following diagram follows by the unitality of the s-action $\varrho_{(tX,\mu_X)}$ and the naturality of the unit η of s.

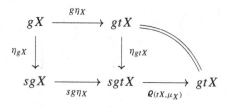

Finally we should show the compatibility of γ with the multiplications of both monads t and s. For this purpose note that the associativity of the multiplication μ of the monad t makes $\mu_X : t^2X \to tX$ a morphism in $\mathbf{A}^t$, from (t^2X, μ_{tX}) to (tX, μ_X), for any object X of **A**. Hence $g\mu_X$ is a morphism in $\mathbf{B}^s$ from $(gt^2X, \varrho_{(t^2X,\mu_{tX})})$ to $(gtX, \varrho_{(tX,\mu_X)})$. Using this, the unitality of μ, the associativity of the action $\varrho_{(tX,\mu_X)}$ and the naturality of μ, we see that the following diagram— expressing the compatibility of γ with the multiplications of t and s—commutes.

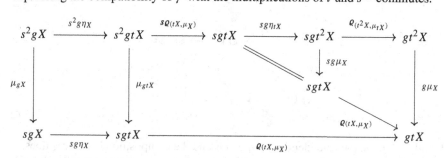

In the opposite direction, assume that the monad morphism γ in part (ii) is given. We construct the desired functor g^γ sending an object (V, v) of $\mathbf{A}^t$ to

$$(gV, \quad sgV \xrightarrow{\ \gamma_V\ } gtV \xrightarrow{\ gv\ } gV\ )$$

and sending a morphism $h : (V, v) \to (V', v')$ to gh; this makes the diagram of part (i) commute.

Unitality of the action $gv \circ \gamma_V : sgV \to gV$ follows by the unitality of v and the compatibility of γ with the units of t and s:

$$
\begin{array}{ccc}
gV & = & gV \\
\eta_{gV} \downarrow & & \downarrow g\eta_V \\
sgV & \xrightarrow{\ \gamma_V\ } gtV & \xrightarrow{\ gv\ } gV
\end{array}
$$

and its associativity follows by the naturality of γ, the associativity of v and the compatibility of γ with the multiplications of t and s:

$$
\begin{array}{ccccc}
s^2gV & \xrightarrow{\ s\gamma_V\ } & sgtV & \xrightarrow{\ sgv\ } & sgV \\
& & \downarrow \gamma_t V & & \downarrow \gamma_V \\
\mu_{gV} \downarrow & & gt^2V & \xrightarrow{\ gtv\ } & gtV \\
& & \downarrow g\mu_V & & \downarrow gv \\
sgV & \xrightarrow{\ \gamma_V\ } & gtV & \xrightarrow{\ gv\ } & gV
\end{array}
$$

This proves that $(gV, gv \circ \gamma_V)$ is an object of $\mathbf{B}^s$. Also gh is a morphism in $\mathbf{B}^s$; that is, the following diagram commutes, by the naturality of γ and since h is a morphism in $\mathbf{A}^t$:

$$
\begin{array}{ccccc}
sgV & \xrightarrow{\ \gamma_V\ } & gtV & \xrightarrow{\ gv\ } & gV \\
sgh \downarrow & & \downarrow gth & & \downarrow gh \\
sgV' & \xrightarrow{\ \gamma_{V'}\ } & gtV' & \xrightarrow{\ gv'\ } & gV'
\end{array}
$$

Finally, g^γ preserves the identity morphisms and the composition because g does.

The above constructions are mutual inverses. Indeed, starting with a functor g^γ as in part (i), and iterating both constructions, we arrive at the functor sending an Eilenberg–Moore t-algebra (V, v) to the pair consisting of the object gV and the s-action occurring in the equal paths around the commutative diagram

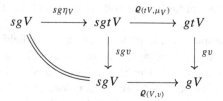

That is, we re-obtain the functor g^γ. Here we used that—by its associativity—v is a morphism in A^t from (tV, μ_V) to (V, v) and hence gv is a morphism in B^s from $(gtV, \varrho_{(tV,\mu_V)})$ to $(gV, \varrho_{(V,v)})$; as well as the unitality of v.

Starting with a monad morphism γ as in part (ii) and iterating both constructions in the opposite order, we arrive at the natural transformation whose component at an arbitrary object X of A occurs in the equal paths around the commutative diagram

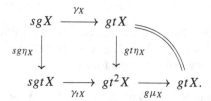

That is, we re-obtain γ. Here we used the naturality of γ and the unitality of μ.   □

Example 2.28.

1. Consider two monads t and t' on the same category A. The liftings of the identity functor $\mathsf{A} \to \mathsf{A}$ to a functor in the top row of the commutative diagram

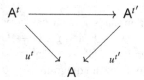

correspond bijectively to the natural transformations $\gamma : t' \to t$ such that $\gamma \circ \mu' = \mu \circ \gamma\gamma$ and $\gamma \circ \eta' = \eta$.

In particular, if the above monads are $t = A \otimes -$ and $t' = A' \otimes -$ on vec, induced by respective algebras A and A' as in Example 2.21 3, then we infer from Exercise 2.8 that the functors $\mathsf{mod}(A) \to \mathsf{mod}(A')$ rendering commutative the diagram

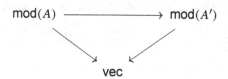

—involving the forgetful functors to vec—correspond bijectively to the algebra homomorphisms $A' \to A$.

2. As in part 4 of Example 2.5, any vector space W over a field k induces a functor $W \otimes - : \text{vec} \to \text{vec}$. We want to lift this functor. To this end, consider the following two monads on vec. The first one is the identity functor as in part 1 of Example 2.21, with Eilenberg–Moore category vec, see part 1 of Example 2.23. The second one is the monad in part 3 of Example 2.21, induced on vec by a k-algebra A, with Eilenberg–Moore category $\text{mod}(A)$ in part 2 of Example 2.23.

We are interested in the lifted functors

$$
\begin{array}{ccc}
\text{vec} & -\,-\,\to & \text{mod}(A) \\
\Big\| & & \Big\downarrow {\scriptstyle u} \\
\text{vec} & \xrightarrow[\;W\otimes-\;]{} & \text{vec}
\end{array}
$$

By Theorem and Definition 2.27, they correspond bijectively to monad morphisms with components $\gamma_X : A \otimes W \otimes X \to W \otimes X$, for any vector space X.

By Exercise 2.8 they must be of the form

$$\gamma_X(a \otimes w \otimes x) = \gamma_k(a \otimes w) \otimes x, \qquad \forall x \in X, \ w \in W, \ a \in A.$$

Moreover, γ obeys the compatibility conditions of a monad morphism if and only if $\gamma_k : A \otimes W \to W$ is a unital and associative A-action on W.

In other words, liftings of the functor $W \otimes - : \text{vec} \to \text{vec}$ to $\text{vec} \to \text{mod}(A)$ are in a bijective correspondence with the A-actions on W.

Exercise 2.29. Consider monads t on a category A, s on B and z on C; together with functors $f : \text{A} \to \text{B}$ and $g : \text{B} \to \text{C}$. Assume that they admit liftings $f^\varphi : \text{A}^t \to \text{B}^s$ and $g^\gamma : \text{B}^s \to \text{C}^z$ along some monad morphisms φ and γ. By the commutativity of the diagram

$$
\begin{array}{ccccc}
\text{A}^t & \xrightarrow{\;f^\varphi\;} & \text{B}^s & \xrightarrow{\;g^\gamma\;} & \text{C}^z \\
{\scriptstyle u^t}\Big\downarrow & & {\scriptstyle u^s}\Big\downarrow & & \Big\downarrow{\scriptstyle u^z} \\
\text{A} & \xrightarrow[\;f\;]{} & \text{B} & \xrightarrow[\;g\;]{} & \text{C}
\end{array}
$$

we know that $g^\gamma f^\varphi$ is a lifting of gf. Compute the corresponding monad morphism.

Theorem and Definition 2.30 ([105]).   For any monads t on a category A and s on a category B, let $h, g : \mathsf{A} \to \mathsf{B}$ be functors admitting liftings $h^\chi, g^\gamma : \mathsf{A}^t \to \mathsf{B}^s$ (along respective monad morphisms χ and γ). Then for any natural transformation $\omega : h \to g$ the following assertions hold.

(1) There exists at most one natural transformation $\overline{\omega} : h^\chi \to g^\gamma$ such that the diagram

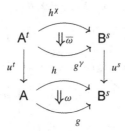

of natural transformations commutes; that is, $u^s \overline{\omega}_{(V,v)} = \omega_V$ for all Eilenberg–Moore t-algebras (V, v).

(2) The natural transformation $\overline{\omega}$ in part (1) exists if and only if the following diagram commutes.

$$
\begin{array}{ccc}
sh & \xrightarrow{\;\chi\;} & ht \\
{\scriptstyle 1\omega}\downarrow & & \downarrow{\scriptstyle \omega 1} \\
sg & \xrightarrow[\;\gamma\;]{} & gt
\end{array}
$$

The natural transformation $\overline{\omega}$—provided that it exists—is called the *lifting* of ω.

Proof. Part (1) is immediate from the faithfulness of u^s.

(2) The lifting $\overline{\omega}$ exists if and only if $\omega_V : hV \to gV$ is a morphism in B^s; that is, the following diagram commutes.

$$
\begin{array}{ccccc}
shV & \xrightarrow{\;\chi V\;} & htV & \xrightarrow{\;hv\;} & hV \\
{\scriptstyle s\omega V}\downarrow & & & & \downarrow{\scriptstyle \omega V} \\
sgV & \xrightarrow[\;\gamma V\;]{} & gtV & \xrightarrow[\;gv\;]{} & gV
\end{array}
\tag{2.4}
$$

If the diagram of part (2) commutes then so does (2.4) by the naturality of ω. Conversely, if (2.4) commutes then so does

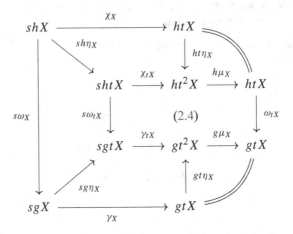

for any object X of A, by the naturality of χ, γ and ω and by the unitality of t.   □

Example 2.31. By part 3 of Example 2.7, any linear map $p : W \to W'$ induces a natural transformation $p \otimes - : W \otimes - \to W' \otimes -$ between functors $\mathsf{vec} \to \mathsf{vec}$. If W and W' are modules over an algebra A, then the induced functors $W \otimes -$ and $W' \otimes - : \mathsf{vec} \to \mathsf{vec}$ lift to $\mathsf{vec} \to \mathsf{mod}(A)$. Theorem and Definition 2.30 says that the natural transformation $p \otimes - : W \otimes - \to W' \otimes -$ between functors $\mathsf{vec} \to \mathsf{vec}$ lifts to a natural transformation between the lifted functors $\mathsf{vec} \to \mathsf{mod}(A)$ if and only if p is an A-module map.

Exercise 2.32. Show that the lifting of natural transformations is compatible with the composition and the Godement product in Paragraph 2.9. More precisely, consider monads t, s and z on respective categories A, B and C. Let k, h and g be functors $\mathsf{A} \to \mathsf{B}$ all of which have liftings k^κ, h^χ and $g^\gamma : \mathsf{A}^t \to \mathsf{B}^s$ (along respective monad morphisms κ, χ and γ) and let h' and g' be functors $\mathsf{B} \to \mathsf{C}$ which have liftings $h'^{\chi'}$ and $g'^{\gamma'} : \mathsf{B}^s \to \mathsf{C}^z$ (along respective monad morphisms χ' and γ'). Verify the following claims.

(1) If both natural transformations $\omega : k \to h$ and $\vartheta : h \to g$ admit liftings $\overline{\omega} : k^\kappa \to h^\chi$ and $\overline{\vartheta} : h^\chi \to g^\gamma$ then the composite $\vartheta \circ \omega : k \to g$ also admits the lifting $\overline{\vartheta} \circ \overline{\omega} : k^\kappa \to g^\gamma$.

(2) If both natural transformations $\vartheta : h \to g$ and $\vartheta' : h' \to g'$ admit liftings $\overline{\vartheta} : h^\chi \to g^\gamma$ and $\overline{\vartheta}' : h'^{\chi'} \to g'^{\gamma'}$ then the Godement product $\vartheta'\vartheta : h'h \to g'g$ also admits the lifting $\overline{\vartheta'\vartheta}$.

Theorem and Definition 2.33 ([30, Theorem 3.13] [69]). Consider a monad t on a category B (with unit η^t and multiplication μ^t) and a monad s on A (with unit η^s and multiplication μ^s). Take an adjunction $l \dashv r : A \to B$ (with unit η and counit ε) such that l admits a lifting $l^\lambda : B^t \to A^s$ along some monad morphism $\lambda : sl \to lt$. The following assertions are equivalent.

(i) r also admits a lifting $r^\varrho : A^s \to B^t$—along some monad morphism ϱ—such that $l^\lambda \dashv r^\varrho$ is an adjunction whose unit is the lifting of η and the counit is the lifting of ε.

(ii) λ is invertible.

In this situation the adjunction $l^\lambda \dashv r^\varrho$ is said to be the *lifting of $l \dashv r$*.

Proof. If r admits some lifting r^ϱ; and η and ε admit liftings $\bar{\eta} : 1 \to r^\varrho l^\lambda$ and $\bar{\varepsilon} : l^\lambda r^\varrho \to 1$, then $\bar{\eta}$ and $\bar{\varepsilon}$ are the unit and the counit of an adjunction $l^\lambda \dashv r^\varrho$, because by Exercise 2.32 the triangle conditions (2.2) on η and ε imply the same conditions for $\bar{\eta}$ and $\bar{\varepsilon}$. Hence we infer from Theorem and Definition 2.27 and Theorem and Definition 2.30 that assertion (i) is equivalent to the existence of a monad morphism $\varrho : tr \to rs$ rendering commutative the following diagrams.

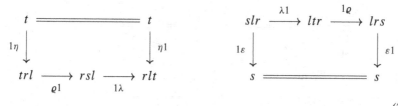

$$(2.5)$$

(The bottom row of the first diagram and the top row of the second diagram are computed as in Exercise 2.29.)

Use the naturality of η and ε together with the triangle conditions (2.2) to see that the adjunction $l \dashv r$ gives rise to mutually inverse bijections between the following families of natural transformations.

$$\Phi \quad : \mathsf{nat}(tr, rs) \to \mathsf{nat}(lt, sl), \; \xi \mapsto \quad lt \xrightarrow{11\eta} ltrl \xrightarrow{1\xi 1} lrsl \xrightarrow{\varepsilon 11} sl$$

$$\Phi^{-1} : \mathsf{nat}(lt, sl) \to \mathsf{nat}(tr, rs), \; \zeta \mapsto \quad tr \xrightarrow{\eta 11} rltr \xrightarrow{1\zeta 1} rslr \xrightarrow{11\varepsilon} rs$$

(ξ and $\Phi(\xi)$ are called *mates* under the adjunction). The conditions of (2.5) are equivalent to the statement that λ and $\Phi(\varrho)$ are mutual inverses. Indeed, apply the functor l to the equal paths of the first diagram of (2.5); and post-compose the resulting equal expressions with $\varepsilon 11$. Use the naturality of ε and a triangle condition from (2.2) to deduce $\lambda \circ \Phi(\varrho) = 1$. Conversely, apply r to both sides of the equality $\lambda \circ \Phi(\varrho) = 1$ and pre-compose the resulting equal expressions with $\eta 1$. Use the naturality of η and a triangle condition from (2.2) to deduce commutativity of the first diagram of (2.5). By symmetric steps commutativity of the second diagram of (2.5) is shown to be equivalent to $\Phi(\varrho) \circ \lambda = 1$.

This shows that whenever (i) holds, λ is invertible as stated in (ii).

Conversely, if λ is invertible as in (ii), then we construct $\varrho := \Phi^{-1}(\lambda^{-1})$. It renders commutative the diagrams of (2.5) by the above considerations and it is a monad morphism by commutativity of the following diagrams (where we use the naturality of the occurring morphisms, the fact that λ is a monad morphism and the triangle conditions (2.2) on η and ε).

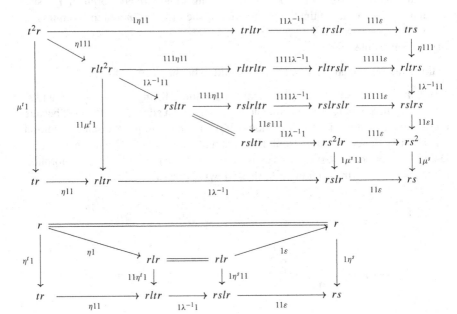

Thus the assertions of (i) hold.   □

Chapter 3
(Hopf) Bimonads

This chapter continues the survey of the theoretical background, turning to more specific topics. Monoidal categories are introduced and monads on them are studied. The general theory of lifting in Chap. 2 is applied to the functors and natural transformations constituting the monoidal structure of the base category. A bijection is proven between the liftings of the monoidal structure of the base category to the Eilenberg–Moore category of a monad; and opmonoidal structures on the monad. Inspired by the examples in the forthcoming chapters, an opmonoidal monad is termed a *bimonad*.

If the base category is closed monoidal, then a sufficient and necessary condition is obtained for the lifting of the closed structure as well, in the form of the invertibility of a canonical natural transformation. Then a *Hopf monad*—on any, not necessarily closed monoidal category—is defined as a bimonad for which this natural transformation is invertible.

The key references here are [30, 79] and [83].

Definition 3.1 ([74, Section VII.1]). A *monoidal structure* on a category A consists of

- a distinguished object I, called the *monoidal unit* (regarded as a functor from the singleton category to A; cf. Example 2.5 6)
- a functor $\otimes$ from the Cartesian product category $A \times A$ (see Example 2.2 10) to A, called the *monoidal product*
- natural isomorphisms $\alpha : (- \otimes -) \otimes - \to - \otimes (- \otimes -)$, called the *associativity constraint*, $\lambda : I \otimes - \to 1$ and $\varrho : - \otimes I \to 1$, called the *left and right unit constraints*, respectively

© Springer Nature Switzerland AG 2018
G. Böhm, *Hopf Algebras and Their Generalizations from a Category Theoretical Point of View*, Lecture Notes in Mathematics 2226,
https://doi.org/10.1007/978-3-319-98137-6_3

such that for all objects X, Y, Z, V, the following diagrams commute.

$$
\begin{array}{ccccc}
((X \otimes Y) \otimes Z) \otimes V & \xrightarrow{\alpha_{X \otimes Y, Z, V}} & (X \otimes Y) \otimes (Z \otimes V) & \xrightarrow{\alpha_{X, Y, Z \otimes V}} & X \otimes (Y \otimes (Z \otimes V)) \\
\alpha_{X, Y, Z} \otimes 1 \downarrow & & & & \uparrow 1 \otimes \alpha_{Y, Z, V} \\
(X \otimes (Y \otimes Z)) \otimes V & & \xrightarrow{\qquad \alpha_{X, Y \otimes Z, V} \qquad} & & X \otimes ((Y \otimes Z) \otimes V)
\end{array}
$$

$$
\begin{array}{ccc}
(X \otimes I) \otimes Y & \xrightarrow{\quad\quad\quad \alpha_{X, I, Y} \quad\quad\quad} & X \otimes (I \otimes Y) \\
& \searrow{\scriptstyle \varrho_X \otimes 1} \qquad \swarrow{\scriptstyle 1 \otimes \lambda_Y} & \\
& X \otimes Y &
\end{array}
$$

They are known as the *pentagon* and *triangle axioms*, respectively.

A monoidal structure is *strict* if its associativity and unit constraints are identity natural isomorphisms.

Example 3.2.

1. The category of sets in Example 2.2 3 is strict monoidal via the monoidal product provided by the Cartesian product and monoidal unit a singleton set.

   More generally, a category **A** is said to have *binary products* if for any pair of objects X and Y there is an object $X \times Y$—the *product object*—together with morphisms $X \xleftarrow{p_X} X \times Y \xrightarrow{p_Y} Y$ which have the following universal property. For any object G and any pair of morphisms $X \xleftarrow{g_X} G \xrightarrow{g_Y} Y$ there is a unique morphism $g : G \to X \times Y$ such that $p_X \circ g = g_X$ and $p_Y \circ g = g_Y$. The best known example of a binary product is the Cartesian product of sets.

   If binary products exist in **A** then any pair of morphisms $f : X \to X'$ and $h : Y \to Y'$ determines a unique morphism $f \times h$ in the commutative diagram

$$
\begin{array}{ccccc}
X & \xleftarrow{\quad p_X \quad} & X \times Y & \xrightarrow{\quad p_Y \quad} & Y \\
f \downarrow & & \downarrow f \times h & & \downarrow h \\
X' & \xleftarrow{\quad p_{X'} \quad} & X' \times Y' & \xrightarrow{\quad p_{Y'} \quad} & Y'.
\end{array}
$$

   A *terminal object* in **A** is an object I to which there is precisely one morphism from any object (like the singleton set in the category of sets).

   If a category has both binary products $\times$ and a terminal object I then it can be seen as a monoidal category $(\mathbf{A}, I, \times)$. Motivated by the Cartesian product of sets, a monoidal category of this form is called a *Cartesian monoidal* category.

2. The category of vector spaces in Example 2.2 4 is monoidal via the monoidal product provided by the tensor product of vector spaces and monoidal unit the base field.

3. For a monoid G, the category vec^G of G-graded vector spaces in Example 2.2 5 admits a monoidal structure. The monoidal product of some objects $\{X_g\}_{g \in G}$ and $\{Y_g\}_{g \in G}$ consists of direct sum vector spaces; it is $\{\oplus_{h \cdot k = g} X_h \otimes Y_k\}_{g \in G}$ with the obvious inducement on the morphisms. The monoidal unit has the base field as the component at the unit of G and the trivial (zero) vector space in every other component.

4. If A is a *commutative* ring then the category $\mathsf{mod}(A)$ of A-modules in Example 2.2 6 is monoidal via the A-module tensor product. The A-module tensor product $V \otimes_A W$ of A-modules V and W is the quotient of the product $V \otimes W$ of Abelian groups,

$$V \otimes_A W := V \otimes W / \{a \cdot v \otimes w - v \otimes a \cdot w \mid a \in A, \ v \in V, \ w \in W\}$$

with the obvious inducement on the morphisms. The monoidal unit is the Abelian group A with the action provided by the multiplication.

5. The category of bimodules over an arbitrary algebra A is monoidal via the monoidal product provided by the A-module tensor product (recalled e.g. in Example 2.5 4) and monoidal unit the regular A-bimodule A with actions given by the multiplication.

6. For any category C, the category $\mathsf{end}(\mathsf{C})$ whose objects are the endofunctors $\mathsf{C} \to \mathsf{C}$, and whose morphisms are the natural transformations, admits the following strict monoidal structure. The monoidal product is the composition of functors (and the induced Godement product of Paragraph 2.9 on the natural transformations). The monoidal unit is the identity functor. The associativity and unit constraints are identity natural transformations.

7. A monoidal structure on a discrete category $\mathsf{disc}(X)$ of Example 2.2 1 consists of an associative and unital multiplication on the set X; that is, the structure of a *monoid*. This monoidal structure is strict.

   Looking at a singleton set as a trivial monoid—consisting only of the unit element—this yields in particular a monoidal structure on the singleton category $\mathbb{1}$ of Example 2.2 1.

8. If H and H' are Hilbert spaces then the tensor product vector space $H \otimes H'$ has an inner product but is not in general complete. A monoidal product on the category hil of Example 2.2 7 is obtained by taking the completion $H \widehat{\otimes} H'$, whose monoidal unit is the one-dimensional Hilbert space of complex numbers—see [65, Propositions 2.6.5 and 2.6.12].

9. The category alg of algebras over a given field in Example 2.2 8 is monoidal via the tensor product of the underlying vector spaces. That is to say, the base field carries a trivial algebra structure and for any algebras X and Y, $X \otimes Y$ also becomes an algebra with the factorwise multiplication $(x \otimes y)(x' \otimes y') = xx' \otimes yy'$ and the unit $1 \otimes 1$. The associativity and unit constraints of vec, as

well as the linear map $f \otimes g$ for any algebra homomorphisms f and g, become algebra homomorphisms for these structures.

10. For any monoidal category $(\mathsf{A}, I, \otimes)$ there is another monoidal category which is the same category A with the reversed monoidal product $V \circledast W := W \otimes V$ and the same monoidal unit I. The left and right unit constraints are

$$I \circledast V = V \otimes I \xrightarrow{\ \varrho_V\ } V \quad \text{and} \quad V \circledast I = I \otimes V \xrightarrow{\ \lambda_V\ } V$$

respectively, and the associativity constraint is

$$(U \circledast V) \circledast W = W \otimes (V \otimes U) \xrightarrow{\ \alpha^{-1}_{W,V,U}\ } (W \otimes V) \otimes U = U \circledast (V \circledast W)$$

for all objects U, V, W.

11. For a monoidal category $(\mathsf{A}, I, \otimes)$, the opposite category of Example 2.2 9. admits the following monoidal structure. The monoidal product and the monoidal unit are

$$\otimes^{\mathrm{op}} : \mathsf{A}^{\mathrm{op}} \times \mathsf{A}^{\mathrm{op}} = (\mathsf{A} \times \mathsf{A})^{\mathrm{op}} \to \mathsf{A}^{\mathrm{op}} \quad \text{and} \quad I^{\mathrm{op}} : \mathbb{1} = \mathbb{1}^{\mathrm{op}} \to \mathsf{A}^{\mathrm{op}},$$

respectively, see Example 2.5 7. The associativity and unit constraints are the inverses of those in $(\mathsf{A}, I, \otimes)$.

12. The Cartesian product $\mathsf{A} \times \mathsf{B}$ in Example 2.2 10, of monoidal categories $(\mathsf{A}, I, \otimes)$ and $(\mathsf{B}, I, \otimes)$, admits the following monoidal structure. The monoidal product is

$$\mathsf{A} \times \mathsf{B} \times \mathsf{A} \times \mathsf{B} \xrightarrow{\ 1 \times \mathrm{flip} \times 1\ } \mathsf{A} \times \mathsf{A} \times \mathsf{B} \times \mathsf{B} \xrightarrow{\ \otimes \times \otimes\ } \mathsf{A} \times \mathsf{B}$$

and the monoidal unit is $I \times I : \mathbb{1} = \mathbb{1} \times \mathbb{1} \to \mathsf{A} \times \mathsf{B}$. The associativity and unit constraints are built up from those in $(\mathsf{A}, I, \otimes)$ and $(\mathsf{B}, I, \otimes)$ in the evident way.

Exercise 3.3. Verify the commutativity of the following diagrams for the associativity and unit constraints of an arbitrary monoidal category $(\mathsf{A}, I, \otimes)$, and arbitrary objects X, Y.

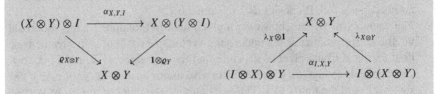

Exercise 3.4. Prove that for the left and right unit constraints λ and ϱ in any monoidal category $(\mathsf{A}, I, \otimes)$, the morphisms λ_I and $\varrho_I : I \otimes I \to I$ are equal.

Definition 3.5 ([74, Section XI.2]). A *monoidal structure* on a functor f from a monoidal category $(\mathsf{A}, I, \otimes)$ to a monoidal category $(\mathsf{A}', I', \otimes')$ consists of

- a morphism $f^0 : I' \to fI$ (called the *nullary part*)
- a natural transformation $f^2 : f - \otimes' f- \to f(- \otimes -)$ (called the *binary part*)

such that the following diagrams commute for any objects X, Y, Z of A.

$$
\begin{array}{ccc}
(fX \otimes' fY) \otimes' fZ & \xrightarrow{\;\alpha'_{fX,fY,fZ}\;} & fX \otimes' (fY \otimes' fZ) \\
\Big\downarrow{\scriptstyle f^2_{X,Y}\otimes'1} & & \Big\downarrow{\scriptstyle 1\otimes' f^2_{Y,Z}} \\
f(X \otimes Y) \otimes' fZ & & fX \otimes' f(Y \otimes Z) \\
\Big\downarrow{\scriptstyle f^2_{X\otimes Y,Z}} & & \Big\downarrow{\scriptstyle f^2_{X,Y\otimes Z}} \\
f((X \otimes Y) \otimes Z) & \xrightarrow[\;f\alpha_{X,Y,Z}\;]{} & f(X \otimes (Y \otimes Z))
\end{array}
$$

$$
\begin{array}{ccc}
I' \otimes' fX & \xrightarrow{\;\lambda'_{fX}\;} & fX \\
\Big\downarrow{\scriptstyle f^0\otimes'1} & & \Big\uparrow{\scriptstyle f\lambda_X} \\
fI \otimes' fX & \xrightarrow[\;f^2_{I,X}\;]{} & f(I \otimes X)
\end{array}
$$

$$
\begin{array}{ccc}
fX \otimes' fI & \xrightarrow{\;\;} & f(X \otimes I) \\
\Big\uparrow{\scriptstyle 1\otimes' f^0} & {\scriptstyle f^2_{X,I}} & \Big\downarrow{\scriptstyle f\varrho_X} \\
fX \otimes' I' & \xrightarrow[\;\varrho'_{fX}\;]{} & fX
\end{array}
$$

The diagram on the left is known as the *associativity condition* and the diagrams on the right expresses the *unitality conditions*. A *monoidal functor* means a functor together with a monoidal structure.

An *opmonoidal structure* on f is a monoidal structure on the functor $f^{\mathrm{op}} : \mathsf{A}^{\mathrm{op}} \to \mathsf{A}'^{\mathrm{op}}$ of Example 2.5 7. That is, it consists of

- a morphism $f^0 : fI \to I'$ (called the *nullary part*)
- a natural transformation $f^2 : f(- \otimes -) \to f - \otimes' f-$ (called the *binary part*)

such that the same diagrams with reversed arrows commute; that is, the *coassociativity* and *counitality* conditions hold. An *opmonoidal functor* means a functor together with an opmonoidal structure.

An (op)monoidal structure is *strong* if the morphism f^0 and the natural transformation f^2 are invertible.

An (op)monoidal structure is *strict* if f^0 is the identity morphism and f^2 is the identity natural transformation. This means the equality of objects $fI = I'$, the equality of functors $f(- \otimes -) = f - \otimes' f-$; and the equalities $\alpha'_{fX,fY,fZ} = f\alpha_{X,Y,Z}$, $\lambda'_{fX} = f\lambda_X$ and $\varrho'_{fX} = f\varrho_X$ on the associativity and unit constraints, for all objects X, Y, Z of A.

Example 3.6.

1. Identity functors on monoidal categories are strict monoidal.
2. The 'linear span' functor in part 2 of Example 2.5 endowed with the obvious nullary and binary parts is strong monoidal.
3. Take an algebra A over a field k. The forgetful functor u (see part 3 of Example 2.5) from the monoidal category of A-bimodules in part 5 of Example 3.2 to the monoidal category of vector spaces in part 2 of Example 3.2, admits the following monoidal structure. The nullary part is the unit of A seen as a linear map $k \to A$, and the binary part $uV \otimes uW \to u(V \otimes_A W)$ is given by the canonical projection, for any A-bimodules V and W. This monoidal structure is not strong.
4. The identity functor vec $\to$ vec can be seen as a strong monoidal functor from the monoidal category in Example 3.2 2 to its reversed form in Example 3.2 10. The nullary part of the monoidal structure is the identity map of the base field. The binary part has the components

$$V \otimes W \to W \otimes V, \qquad v \otimes w \mapsto w \otimes v$$

   provided by the flip maps, for any vector spaces V and W.
5. From any monoidal category $(\mathsf{A}, I, \otimes)$ there is a strong monoidal functor to the monoidal category end(A) of Example 3.2 6. It sends an object X to the *induced functor* $X \otimes - : \mathsf{A} \to \mathsf{A}$ and it sends a morphism $f : X \to Y$ to the natural transformation $f \otimes 1 : X \otimes - \to Y \otimes -$. The nullary part of its opmonoidal structure is given by the left unit constraint $I \otimes - \to 1$ and the binary part is given by the associativity constraint $(X \otimes Y) \otimes - \to X \otimes (Y \otimes -)$ for all objects X, Y of A.
6. Consider a monoidal functor (f, f^0, f^2) and a natural isomorphism $\varphi : f \to g$. There is a monoidal structure on g with nullary and binary parts

$$I \xrightarrow{f^0} fI \xrightarrow{\varphi_I} gI \qquad gX \otimes gY \xrightarrow{\varphi_X^{-1} \otimes \varphi_Y^{-1}} fX \otimes fY \xrightarrow{f_{X,Y}^2} f(X \otimes Y) \xrightarrow{\varphi_{X \otimes Y}} g(X \otimes Y).$$

Lemma 3.7 ([62, Section 1]). *From any monoidal category $(\mathsf{A}, I, \otimes)$ there is a strong monoidal equivalence to the following strict monoidal category. The objects consist of a functor $t : \mathsf{A} \to \mathsf{A}$ and a natural isomorphism $\tau : (t-) \otimes - \to t(-\otimes -)$ such that for all objects X, Y, Z the following diagram commutes.*

$$
\begin{array}{ccc}
(tX \otimes Y) \otimes Z & \xrightarrow{\tau_{X,Y} \otimes 1} \ t(X \otimes Y) \otimes Z \xrightarrow{\tau_{X \otimes Y, Z}} & t((X \otimes Y) \otimes Z) \\
{\scriptstyle \alpha_{tX,Y,Z}} \downarrow & & \downarrow {\scriptstyle t\alpha_{X,Y,Z}} \\
tX \otimes (Y \otimes Z) & \xrightarrow[\tau_{X, Y \otimes Z}]{\hspace{5cm}} & t(X \otimes (Y \otimes Z))
\end{array}
\tag{3.1}
$$

The morphisms $(t, \tau) \to (t', \tau')$ are natural transformations $\varphi : t \to t'$ such that for all objects X, Y of A *the following diagram commutes.*

$$
\begin{array}{ccc}
tX \otimes Y & \xrightarrow{\ \tau_{X,Y}\ } & t(X \otimes Y) \\
\varphi_X \otimes 1 \downarrow & & \downarrow \varphi_{X \otimes Y} \\
t'X \otimes Y & \xrightarrow[\ \tau'_{X,Y}\ ]{} & t'(X \otimes Y)
\end{array}
\tag{3.2}
$$

The monoidal product of the objects (t, τ) and (t', τ') consists of the composite functor $t't$ and the natural isomorphism whose components are

$$
t'tX \otimes Y \xrightarrow{\ \tau'_{tX,Y}\ } t'(tX \otimes Y) \xrightarrow{\ t'\tau_{X,Y}\ } t't(X \otimes Y) \ .
$$

The monoidal product of morphisms is the Godement product of natural transformations. The monoidal unit is the identity functor with the identity natural isomorphism.

Proof. A functor l from A to the stated category sends an object X to the pair consisting of the induced functor $X \otimes - : A \to A$ and the natural isomorphism $\alpha_{X,-,-} : (X \otimes -) \otimes - \to X \otimes (- \otimes -)$ provided by the associativity constraint. It renders commutative (3.1) by the pentagon axiom. The functor l sends a morphism $f : X \to Y$ to the natural transformation $f \otimes 1 : X \otimes - \to Y \otimes -$. It renders commutative (3.2) by the naturality of α. The nullary part of its opmonoidal structure is given by the left unit constraint λ and the binary part is given by the associativity constraint α.

In the opposite direction an object (t, τ) is sent to the object tI of A; and a morphism $\varphi : (t, \tau) \to (t', \tau')$ is sent to $\varphi_I : tI \to t'I$.

Starting with an object X of A and applying both functors above we obtain $X \otimes I$. So a natural isomorphism from the composite functor to the identity functor is given by the right unit constraint.

Starting with an object (t, τ) in the other category and applying the above functors in the opposite order we obtain the object $(tI \otimes -, \tau_I \otimes 1)$. So a natural isomorphism from the composite functor to the identity functor is given by the components

$$
tI \otimes X \xrightarrow{\ \tau_{I,X}\ } t(I \otimes X) \xrightarrow{\ t\lambda_X\ } tX.
$$

It makes (3.2) commute by the commutativity of

$$
\begin{array}{ccc}
(tI \otimes X) \otimes Y & \xrightarrow{\ \tau_{I,X} \otimes 1\ } t(I \otimes X) \otimes Y & \xrightarrow{\ t\lambda_X \otimes 1\ } tX \otimes Y \\
\end{array}
$$

(with $\alpha_{tI,X,Y}$, $\tau_{I \otimes X, Y}$, $\tau_{X,Y}$, (3.1), $t((I \otimes X) \otimes Y)$, $t(\lambda_X \otimes 1)$, $t\alpha_{I,X,Y}$, Exercise 3.3, and the bottom row)

$$
tI \otimes (X \otimes Y) \xrightarrow{\ \tau_{I,X \otimes Y}\ } t(I \otimes (X \otimes Y)) \xrightarrow{\ t\lambda_{X \otimes Y}\ } t(X \otimes Y).
$$

$\square$

3.8. Coherence. Mac Lane's Coherence Theorem [62, Section 1] and [74, Section VII.2] asserts—somewhat roughly speaking—that Lemma 3.7 allows us to prove that all coherence diagrams—those which are built up only from the associativity and unit constraints, taking possibly their inverses and monoidal products with identity morphisms; see e.g. the diagrams of Exercises 3.3 and 3.4—commute.

Consequently—and since we are only interested, on the other hand, in structures and properties which are preserved by strong monoidal equivalences; see e.g. Examples 3.18 3 and 3.21 3 below—, for brevity, often we shall not explicitly denote the associativity and the unit constraints in a monoidal category. That is, we shall write X instead of both monoidal products $I \otimes X$ and $X \otimes I$ with the unit object I; and we write $X \otimes Y \otimes Z$ instead of both $(X \otimes Y) \otimes Z$ and $X \otimes (Y \otimes Z)$, for any objects X, Y, Z.

Exercise 3.9. Consider equivalence functors $f : A \to B$ and $g : B \to A$ and a strict monoidal structure $(I, \otimes)$ on A. Construct a monoidal structure on B with respect to which f is strong (but not necessarily strict!) monoidal.

Exercise 3.10. Show that the composite of monoidal functors is monoidal; symmetrically, the composite of opmonoidal functors is opmonoidal.

Exercise 3.11. Prove that in an adjunction $l \dashv r$ between monoidal categories, there is a bijective correspondence between the monoidal structures on r and the opmonoidal structures on l.

Definition 3.12 ([30, Section 1.2] and [79, p. 472]). A natural transformation $\varphi : f \to f'$ between monoidal functors is said to be *monoidal* if the following diagrams commute.

$$
\begin{array}{ccc}
fX \otimes fY & \xrightarrow{\ f^2_{X,Y}\ } & f(X \otimes Y) \\
\varphi_X \otimes \varphi_Y \downarrow & & \downarrow \varphi_{X \otimes Y} \\
f'X \otimes f'Y & \xrightarrow[\ f'^2_{X,Y}\ ]{} & f'(X \otimes Y)
\end{array}
\qquad
\begin{array}{ccc}
I & \xrightarrow{\ f^0\ } & fI \\
\| & & \downarrow \varphi_I \\
I & \xrightarrow[\ f'^0\ ]{} & f'I
\end{array}
$$

A natural transformation $\varphi : f \to f'$ between opmonoidal functors is said to be *opmonoidal* if it is a monoidal natural transformation $f'^{\mathrm{op}} \to f^{\mathrm{op}}$ (see Example 2.5 7); that is, the same diagrams with reversed horizontal arrows commute. (So if both f and f' are strong monoidal, then a natural transformation $f \to f'$ is monoidal if and only if it is opmonoidal.)

Example 3.13. Identity natural transformations of monoidal functors are monoidal and identity natural transformations of opmonoidal functors are opmonoidal.

Exercise 3.14. Prove the isomorphism (in the sense of Example 2.13 1) of the following categories.

- The category **alg** of algebras over a given field k in Example 2.2 8.
- The category whose objects are the monoidal functors from the monoidal singleton category $\mathbb{1}$ of Example 3.2 7 to the monoidal category **vec** of k-vector spaces in Example 3.2 2; and whose morphisms are the monoidal natural transformations.

Symmetrically, prove that the category of coalgebras is isomorphic to the category of opmonoidal functors $\mathbb{1} \to$ **vec**.

Exercise 3.15. Consider an adjunction $l \dashv r$ and a strong monoidal structure (r^0, r^2) on r; then there is a corresponding opmonoidal structure (l^0, l^2) on l in Exercise 3.11. Regard on r the opmonoidal structure provided by the inverses of r^0 and r^2; and on the composite functors rl and lr take the opmonoidal structures in Exercise 3.10. Prove that with respect to these structures the unit and the counit of the adjunction are opmonoidal natural transformations.

Exercise 3.16. Show that both the composite and the Godement product in Paragraph 2.9 of monoidal natural transformations is monoidal (regarding the monoidal structure of the composite functors in Exercise 3.10). Consequently the analogous statement holds for opmonoidal natural transformations too.

Definition 3.17 ([30, Section 2.4] and [79, Example 2.5]). An *opmonoidal monad* on a monoidal category A consists of

- a monad t on the category A (with multiplication μ and unit η)
- an opmonoidal structure $(t^0 : tI \to I, t^2 : t(- \otimes -) \to t - \otimes t-)$ on the functor t

such that μ and η are opmonoidal natural transformations; that is, the following diagrams commute. (The morphisms in the top row of the first diagram in each row are computed as in Exercise 3.10.)

$$
\begin{array}{ccccc}
tt(X \otimes Y) & \xrightarrow{tt^2_{X,Y}} & t(tX \otimes tY) & \xrightarrow{t^2_{tX,tY}} & ttX \otimes ttY \\
\downarrow{\scriptstyle \mu_{X\otimes Y}} & & & & \downarrow{\scriptstyle \mu_X \otimes \mu_Y} \\
t(X \otimes Y) & & \xrightarrow{\quad\quad t^2_{X,Y} \quad\quad} & & tX \otimes tY
\end{array}
\qquad
\begin{array}{ccc}
X \otimes Y & =\!=\!= & X \otimes Y \\
\downarrow{\scriptstyle \eta_{X\otimes Y}} & & \downarrow{\scriptstyle \eta_X \otimes \eta_Y} \\
t(X \otimes Y) & \xrightarrow{t^2_{X,Y}} & tX \otimes tY
\end{array}
$$

$$
\begin{array}{ccccc}
ttI & \xrightarrow{tt^0} & tI & \xrightarrow{t^0} & I \\
\downarrow{\scriptstyle \mu_I} & & & & \| \\
tI & & \xrightarrow{\quad t^0 \quad} & & I
\end{array}
\qquad
\begin{array}{ccc}
I & =\!=\!= & I \\
\downarrow{\scriptstyle \eta_I} & & \| \\
tI & \xrightarrow{t^0} & I
\end{array}
$$

Opmonoidal monads are also called *bimonads*.

Example 3.18.

1. Any identity functor with its trivial monad structure in Example 2.21 1 and with the trivial opmonoidal structure as in Example 3.6 1 is an opmonoidal monad.
2. Consider an opmonoidal monad t and a natural isomorphism $t \to s$. With the monad structure in Example 2.21 4 and with the opmonoidal structure as in Example 3.6 6, s is an opmonoidal monad.
3. By Example 2.21 2, any adjunction $l \dashv r : \mathsf{B} \to \mathsf{A}$ and any monad t on B determine a monad rtl on A.

    If in addition r is strong monoidal, then l is opmonoidal by Exercise 3.11. Since r is opmonoidal too, the composite functor $rtl : \mathsf{A} \to \mathsf{A}$ is opmonoidal by Exercise 3.10 for any opmonoidal functor $t : \mathsf{B} \to \mathsf{B}$. Moreover, the unit and the counit of the adjunction $l \dashv r$ are opmonoidal natural transformations by Exercise 3.15. Hence any opmonoidal monad t on B induces an opmonoidal monad rtl on A by Exercise 3.16.

There are two cases of particular interest. If we choose t above to be the identity opmonoidal monad, then we obtain an opmonoidal monad rl from any adjunction $l \dashv r$ in which r is strong monoidal.

On the other hand, for an equivalence with functors $f : A \to B$ and $g : B \to A$, f is the left adjoint of g by Exercise 2.19. So whenever g is strong monoidal, we obtain an opmonoidal monad gtf on A from any opmonoidal monad t on B.

Theorem 3.19 ([79, Corollary 3.13] and [83, Theorem 7.1]). *Consider a monoidal category* $(A, I, \otimes)$ *and a monad t on the category* A *(with multiplication μ and unit η). There is a bijective correspondence between the following data.*

(i) Monoidal structures $(\overline{I}, \overline{\otimes})$ *on the Eilenberg–Moore category* A^t *such that the forgetful functor* $u^t : A^t \to A$ *is strict monoidal.*

(ii) Liftings of the functors $I : \mathbb{1} \to A$ *and* $\otimes : A \times A \to A$ *as in*

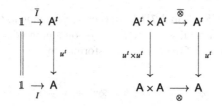

together with liftings of the natural transformations $\alpha : \otimes(\otimes \times 1) \to \otimes(1 \times \otimes)$, $\lambda : \otimes(I \times 1) \to 1$ *and* $\varrho : \otimes(1 \times I) \to 1$ *as in*

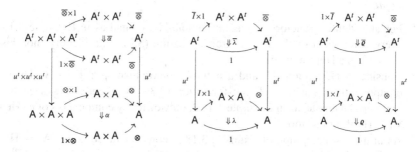

(iii) Opmonoidal monad structures on t.

Proof. Strict monoidality of the functor u^t in part (i) implies that the monoidal unit $\overline{I}$, the monoidal product $\overline{\otimes}$ and the associativity and unit constraints $\overline{\alpha}$, $\overline{\lambda}$ and $\overline{\varrho}$ of A^t are liftings of the respective data I, $\otimes$, α, λ and ϱ in A, as required in (ii).

Conversely, the lifted data in part (ii) satisfy the pentagon and triangle axioms of Definition 3.1 by Exercise 2.32. Hence they provide a monoidal structure on A^t as in (i).

By Theorem and Definition 2.27, the liftings of the functors $I : \mathbb{1} \to \mathsf{A}$ and $\otimes : \mathsf{A} \times \mathsf{A} \to \mathsf{A}$ to respective functors $\overline{I} : \mathbb{1} \to \mathsf{A}^t$ and $\overline{\otimes} : \mathsf{A}^t \times \mathsf{A}^t \to \mathsf{A}^t$ as in part (ii) correspond bijectively to monad morphisms $t^0 : tI \to I$ and $t^2 : t(- \otimes -) \to t - \otimes t-$, respectively. The diagrams expressing that t^0 and t^2 are monad morphisms are identical to the diagrams appearing in Definition 3.17.

By Theorem and Definition 2.30, the natural transformations α, λ and ϱ admit liftings as in part (ii) if and only if they satisfy certain compatibility conditions with t^0 and t^2. Spelling out these compatibility conditions, we get precisely the coassociativity and counitality diagrams in Definition 3.5 defining an opmonoidal structure (t^0, t^2) on the functor t. Together with the diagrams of the previous paragraph this says that (t^0, t^2) equip t with the structure of an opmonoidal monad as in part (iii). $\qquad\qquad\square$

Note that the implication (i)⇒(iii) of Theorem 3.19 also follows alternatively by Exercise 3.15: if the right adjoint functor u^t in part (i) is strict monoidal, then with its left adjoint f^t in Paragraph 2.24 they induce the opmonoidal monad $u^t f^t = t$.

Definition 3.20 ([30, Section 2.7]). An opmonoidal monad t—with monad structure (η, μ) and opmonoidal structure (t^0, t^2)—on a monoidal category $(\mathsf{A}, I, \otimes)$ is said to be a *Hopf monad* if the natural transformation

$$t(- \otimes t-) \ \xrightarrow{\ t^2\ } \ t - \otimes tt - \ \xrightarrow{\ 1 \otimes \mu_-\ } \ t - \otimes t- \tag{3.3}$$

(between functors $\mathsf{A} \times \mathsf{A} \to \mathsf{A}$) is invertible.

Example 3.21.

1. For any identity functor seen as an opmonoidal monad as in Example 3.18 1, every component of the natural transformation (3.3) is an identity morphism. Hence it is a Hopf monad.
2. Consider a Hopf monad t and a natural isomorphism $\varphi : t \to s$. Then the components of the natural transformation (3.3) for t; and for s seen as an opmonoidal monad as in Example 3.18 2, only differ by components of φ. Hence s is a Hopf monad too.
3. As in the last paragraph of Example 3.18 3, consider functors $f : \mathsf{A} \to \mathsf{B}$ and $g : \mathsf{B} \to \mathsf{A}$ which constitute an equivalence, and opmonoidal structure (g^0, g^2) on g which is strong. Then the inverses of g^0 and g^2 yield a monoidal structure. Seeing f as the left adjoint of g as in Exercise 2.19, there is a corresponding opmonoidal structure (f^0, f^2) on f as in Exercise 3.11 which is evidently strong too. By Example 3.18 3 any opmonoidal monad t on B induces an opmonoidal monad gtf on A. The component of the natural transformation (3.3) for this opmonoidal monad, at arbitrary objects X and Y of A, occurs in the equal paths

of the following commutative diagram.

$$gtf(X \otimes gtfY)$$

Since f and g are strong monoidal, the first and the last bits of the left-bottom path are isomorphisms. Since f and g constitute an equivalence, so is the second bit by Exercise 2.19. Whenever t is a Hopf monad, the penultimate bit is also an isomorphism, proving that gtf is a Hopf monad too.

In particular, choosing t above to be the identity Hopf monad, we obtain a Hopf monad gf from any strong monoidal functors f and g which constitute an equivalence.

Remark 3.22. To be fully precise, what we called in Definition 3.20 a Hopf monad, is called in [30, Section 2.7] a *left Hopf monad*. There also a symmetric notion, so-called *right Hopf monad* is introduced, as a bimonad for which the natural transformation

$$t(t - \otimes -) \xrightarrow{t^2} tt - \otimes t - \xrightarrow{\mu_- \otimes 1} t - \otimes t - \qquad (3.4)$$

(between functors $A \times A \to A$) is invertible. In the terminology of [30, Section 2.7] a *Hopf monad* is a bimonad which is both a left and right Hopf monad.

Here we use a somewhat different terminology justified by Theorem 4.8—which relates Hopf monads in the sense of Definition 3.20 to Hopf algebras; see Remark 4.10.

Definition 3.20 does not seem to have any transparent meaning for an opmonoidal monad on an arbitrary monoidal category. It has so, however, for a distinguished class of monoidal categories to be discussed next.

Definition 3.23 ([30, Section 3.1] and [51]). A monoidal category A is said to be *left closed* if for any object X, the functor $- \otimes X : A \to A$ possesses a right adjoint. The right adjoint is called the *internal hom* functor and it is denoted by $[X, -]$. The unit and the counit of the adjunction $- \otimes X \dashv [X, -] : A \to A$ will be denoted by η^X and ε^X, respectively.

Example 3.24.

1. The monoidal category of sets is left closed, with internal hom functor $[X, -]$ sending a set Y to the set of maps $X \to Y$; and sending a map $f : Y \to Y'$ to the post-composition map $f \circ - : [X, Y] \to [X, Y']$.
2. The monoidal category of vector spaces is left closed, with internal hom functor $[X, -]$ sending a vector space Y to the vector space of linear maps $X \to Y$ (with the pointwise linear structure) and sending a linear map $Y \to Y'$ to the post-composition linear map $f \circ - : [X, Y] \to [X, Y']$.
3. The monoidal category of bimodules over an arbitrary algebra A is left closed, with internal hom functor $[X, -]$ sending an A-bimodule Y to the A-bimodule of right A-module maps $X \to Y$ (where the left and right A-actions on a right A-module map $h : X \to Y$ are given by $(a \cdot h \cdot a')x := a \cdot h(a' \cdot x))$; and sending an A-bimodule map $Y \to Y'$ to the post-composition A-bimodule map $f \circ - : [X, Y] \to [X, Y']$.

Definition 3.25 ([30, Section 3.2]). A strict monoidal functor $u : \mathsf{A}' \to \mathsf{A}$ between left closed monoidal categories *strictly preserves* the left closed structure if

$$u[X, -]' \xrightarrow{\eta^{uX}_{u[X, -]'}} [uX, u[X, -]' \otimes uX] = [uX, u([X, -]' \otimes' X)] \xrightarrow{[uX, u\varepsilon'^X]} [uX, u-]$$

(3.5)

is the identity natural transformation for all objects X of A'.

Lemma 3.26. *For a strict monoidal functor* $u : \mathsf{A}' \to \mathsf{A}$ *between left closed categories, the following assertions are equivalent.*

(i) *It strictly preserves the left closed structure; that is, (3.5) is equal to the identity natural transformation.*

(ii) *The functor equality* $u[-, -]' = [u-, u-]$ *holds and the following diagram commutes for any objects* X, Y *of* A'.

$$
\begin{array}{ccc}
uY & \xrightarrow{\eta^{uX}_{uY}} & [uX, uY \otimes uX] \\
{\scriptstyle u\eta'^X_Y} \downarrow & & \| \\
u[X, Y \otimes' X]' & = & [uX, u(Y \otimes' X)]
\end{array}
$$

(3.6)

(iii) *The functor equality* $u[-, -]' = [u-, u-]$ *holds and the following diagram commutes for any objects* X, Y *of* A'.

$$
\begin{array}{ccc}
u[X, Y]' \otimes uX & = & u([X, Y]' \otimes' X) \\
\| & & \downarrow {\scriptstyle u\varepsilon'^X_Y} \\
[uX, uY] \otimes uX & \xrightarrow[\varepsilon^{uX}_{uY}]{} & uY
\end{array}
$$

(3.7)

Proof. The adjunction $- \otimes' X \dashv [X, -]'$ gives rise to a bijection between natural transformations $u[X, -]' \to [uX, u-]$ and $u \to [uX, u(- \otimes' X)]$ as in Exercise 2.18 (1). Both sides of the equalities of natural transformations in parts (i) and (ii) are related by this.

Symmetrically, the adjunction $- \otimes uX \dashv [uX, -]$ gives rise to a bijection between natural transformations $u[X, -]' \to [uX, u-]$ and $u[X, -]' \otimes' uX \to u$ as in Exercise 2.18 (2). Both sides of the equalities of natural transformations in parts (i) and (iii) are related by that. □

Theorem 3.27 ([30, Theorem 3.6]). *For an opmonoidal monad (a.k.a. bimonad) t (with monad structure (η^t, μ^t) and opmonoidal structure (t^0, t^2)) on a left closed monoidal category* A *the following assertions are equivalent.*

(i) A^t *is a left closed monoidal category and the forgetful functor u^t strictly preserves the left closed structure.*
(ii) *For any Eilenberg–Moore t-algebra (V, v), the adjunction $- \otimes V \dashv [V, -]$:* A $\to$ A *admits a lifting $- \otimes (V, v) \dashv [(V, v), -]'$:* A^t $\to$ A^t *along the forgetful functor u^t :* A^t $\to$ A *(in the sense of Theorem and Definition 2.33).*
(iii) *The bimonad t is a Hopf monad.*

Proof. Since the forgetful functor u^t : A^t $\to$ A is strict monoidal, the functor $- \otimes (V, v)$: A^t $\to$ A^t is a lifting of $- \otimes V$: A $\to$ A for any Eilenberg–Moore t-algebra (V, v), along the monad morphism

$$\beta_{-,(V,v)} := \quad t(- \otimes V) \xrightarrow{\; t^2_{-,V} \;} t - \otimes t V \xrightarrow{\; 1 \otimes v \;} t - \otimes V. \tag{3.8}$$

The assertion in part (ii) boils down to three statements about the lifting of

(·) the functor $[V, -]$: A $\to$ A to $[(V, v), -]'$: A^t $\to$ A^t
(··) the natural transformation $\eta^V : 1 \to [V, - \otimes V]$ to $\eta^{(V,v)} : 1 \to [(V, v), - \otimes (V, v)]'$
(∴) the natural transformation $\varepsilon^V : [V, -] \otimes V \to 1$ to $\varepsilon^{(V,v)} : [(V, v), -]' \otimes (V, v) \to 1$

for any Eilenberg–Moore t-algebra (V, v). If they hold then $\eta^{(V,v)}$ is the unit and $\varepsilon^{(V,v)}$ is the counit of the adjunction $- \otimes (V, v) \dashv [(V, v), -]'$: A^t $\to$ A^t by Exercise 2.32. By Lemma 3.26 (·) and (··) hold if and only if (·) and (∴) hold if and only if the assertion of part (i) in the claim holds.

By Theorem and Definition 2.33, (ii) is equivalent to the invertibility of $\beta_{-,(V,v)}$ of (3.8) for all objects (V, v) of A^t.

If $\beta_{-,(V,v)}$ is invertible for all objects (V, v) of A^t, then in particular $\beta_{X, f^t Y}$—that is, the natural transformation of (3.3)—is invertible for all objects X, Y of A (where f^t is the left adjoint of u^t from Paragraph 2.24). This proves (ii)⟹(iii).

Conversely, assume that (iii) holds; that is, $\beta_{X, f^t Y}$ is invertible for all objects X, Y of A. Then since via the t-module epimorphism $v : f^t V = (tV, \mu^t_V) \to (V, v)$ any Eilenberg–Moore t-algebra (V, v) is the quotient of a free one $f^t V$,

$\beta_{X,(V,v)}$ also possesses the inverse

$$tX \otimes V \xrightarrow{\;1\otimes\eta_V^t\;} tX \otimes tV \xrightarrow{\;\beta_{X,f^tV}^{-1}\;} t(X \otimes tV) \xrightarrow{\;t(1\otimes v)\;} t(X \otimes V).$$

It is seen to be the two-sided inverse of $\beta_{X,(V,v)}$ by the commutativity of the following diagrams. (Here we use the naturality of η^t and of $\beta_{-,(V,v)}$ in (V, v), and the unitality and associativity of v.)

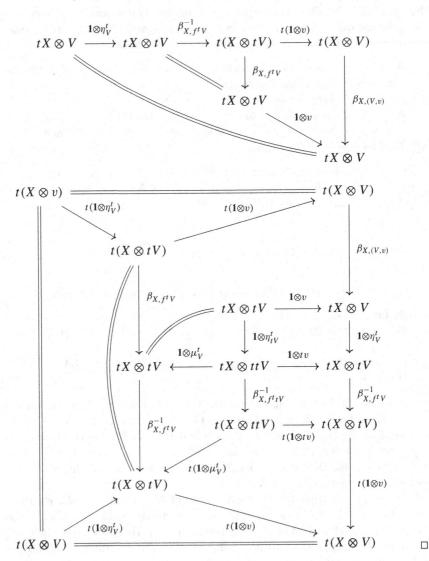

3.28. Alternative Notions of (Hopf) Bimonad. The reader should be warned that in [83, Definition 1.1] Ieke Moerdijk used the name *Hopf monad* for an opmonoidal monad. This is not consistent with our terminology in Theorem 3.19 and Definition 3.20.

On the other hand, in [81, Definition 4.1] Bachuki Masablishvili and Robert Wisbauer used the same phrase *bimonad* in a completely different sense. In that paper, a bimonad means an endofunctor t on an arbitrary, not necessarily monoidal category. It not only carries a monad structure (with associative multiplication $\mu : t^2 \to t$ and unit $\eta : 1 \to t$) but also a *comonad* structure with coassociative *comultiplication* $\delta : t \to t^2$ and *counit* $\varepsilon : t \to 1$.

In order to formulate the compatibility axioms between the monad and comonad structures, the existence of a *mixed distributive law* $\gamma : t^2 \to t^2$ is assumed. This means a monad morphism as in part (ii) of Theorem and Definition 2.27 which is a comonad morphism as well in the sense of the following commutative diagrams.

$$
\begin{array}{ccc}
t^2 & \xrightarrow{\quad\gamma\quad} & t^2 \\
\scriptstyle 1\delta \downarrow & & \downarrow \scriptstyle \delta 1 \\
t^3 \xrightarrow{\;\gamma 1\;} t^3 \xrightarrow{\;1\gamma\;} & & t^3
\end{array}
\qquad
\begin{array}{ccc}
t^2 & \xrightarrow{\;\gamma\;} & t^2 \\
\scriptstyle 1\varepsilon \downarrow & & \downarrow \scriptstyle \varepsilon 1 \\
t & = & t.
\end{array}
$$

With its help the bimonad axioms of [81, Definition 4.1] are encoded in the commutative diagrams

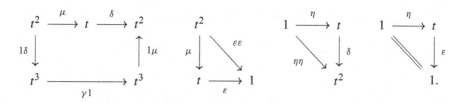

As comonad structures and opmonoidal structures on a functor are in general unrelated, there is no general relation between Mesablishvili–Wisbauer bimonads and bimonads in the sense of Definition 3.17.

In [81, Paragraph 5.5] a Mesablishvili–Wisbauer bimonad is called a *Hopf monad* if the natural transformation $t^2 \xrightarrow{\delta 1} t^3 \xrightarrow{1\mu} t^2$; equivalently, $t^2 \xrightarrow{1\delta} t^3 \xrightarrow{\mu 1} t^2$ is invertible. In this case t comes equipped with an *antipode* natural transformation $\sigma : t \to t$ which renders commutative the diagram

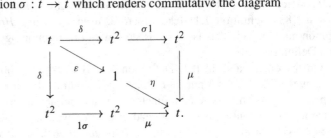

Chapter 4
(Hopf) Bialgebras

The program—of interpreting Hopf algebras and their various generalizations as Hopf monads—begins in this chapter with classical Hopf algebras over fields.

Endofunctors $A \otimes -$ on the category of vector spaces are considered, which are induced by taking the tensor product with a fixed vector space A; see part 4 of Example 2.5. The algebra structures on A are related to the monad structures on the induced functor $A \otimes -$.

The monoidal structure of the category vec of vector spaces (over some field k) was discussed in part 2 of Example 3.2. The opmonoidal structures with respect to it, on an endofunctor of the form $A \otimes -$ like above, are related to the coalgebra structures on the vector space A. This results in a bijection between the *bialgebras* A; and the bimonads with underlying functor $A \otimes -$ on the category of vector spaces. The bijection is shown to restrict to *Hopf algebras* on one hand; and Hopf monads on the other.

The most important references are [101, 102, 109] and [110].

Proposition 4.1. *For any vector space A over a field k, there is a bijective correspondence between the following structures.*

(i) Monad structures on the functor $A \otimes - :$ vec $\to$ vec (see part 4 of Example 2.5).

(ii) Algebra structures on the vector space A.

Furthermore, in this setting the Eilenberg–Moore category of the monad $A \otimes - :$ vec $\to$ vec in part (i) is the category mod(A) of left modules over the algebra A in part (ii).

Proof. For any vector spaces V and W the equivalence of Lemma 3.7 induces a bijection between the linear maps $V \to W$ and those natural transformations

© Springer Nature Switzerland AG 2018
G. Böhm, *Hopf Algebras and Their Generalizations from a Category Theoretical Point of View*, Lecture Notes in Mathematics 2226,
https://doi.org/10.1007/978-3-319-98137-6_4

$\varphi : V \otimes - \to W \otimes -$ between the induced functors $\mathsf{vec} \to \mathsf{vec}$ which render commutative the diagram of (3.2) replacing the horizontal arrows by the associativity constraint in vec.

Now Exercise 2.8 says that any natural transformation $\varphi : V \otimes - \to W \otimes -$ is determined by its component at the base field k as $\varphi = \varphi_k \otimes \mathbf{1}$. Using this form of φ it is easily seen to make (3.2) commute. Thus from Lemma 3.7 we obtain a bijection between the linear maps $V \to W$ and all natural transformations $\varphi : V \otimes - \to W \otimes -$.

This yields the stated bijection between the natural transformations $\mu : A \otimes A \otimes - \to A \otimes -$ and the linear maps $\mu_k : A \otimes A \to A$; and also between the natural transformations $\eta : 1 \to A \otimes -$ and the linear maps $\eta_k : k \to A$. Using the explicit form of these correspondences, the associativity and unitality axioms for any monad $(A \otimes -, \eta, \mu)$ can be formulated as the commutativity of the following diagrams for any vector space V.

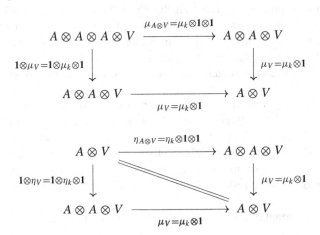

They commute for any vector space V if and only if they commute for $V = k$; that is, μ_k is an associative multiplication with the unit η_k rendering the vector space A an algebra.

The final claim can be found in part 2 of Example 2.23.   □

4.2. The Sweedler–Heynemann Index Notation.

An *algebra* structure on a vector space A over a field k consists of an associative *multiplication*, which is a linear map $A \otimes A \to A$, and a *unit* which can be seen as a linear map $k \to A$. The multiplication sends a rank one tensor $a \otimes a'$ to an element of A, usually denoted as aa'.

Dually, a *coalgebra* structure on A is provided by a coassociative *comultiplication* which is a linear map $A \to A \otimes A$, and a *counit* linear map $A \to k$. In general the comultiplication sends an element a of A to a finite sum of rank one tensors. Writing out all of the occurring summation symbols would result in heavy formulae that would be difficult to read. For this reason the following simplified notation was suggested by Sweedler and Heynemann, see [108, Section 1.2].

Omitting the summation symbol, denote the image of an element a of a coalgebra A under the comultiplication simply by $a_1 \otimes a_2$. But *do keep in mind* that these indices 1 and 2 are to remind you of the summation which is not explicitly denoted.

The counit is a linear map $e : A \to k$ which can be applied to either factor of $a_1 \otimes a_2$ for any element a of A, yielding an element of A. The *counit axioms* of coalgebra in the Sweedler–Heynemann implicit summation convention say that $e(a_1)a_2 = a = a_1 e(a_2)$, for all $a \in A$.

Also the comultiplication is a linear map which can be applied to either factor of $a_1 \otimes a_2$, yielding the respective elements $a_{11} \otimes a_{12} \otimes a_2$ and $a_1 \otimes a_{21} \otimes a_{22}$ of $A \otimes A \otimes A$, where now double summations are understood. The *coassociativity axiom* of coalgebra says that these sums are equal. So by the Sweedler–Heynemann convention we write $a_1 \otimes a_2 \otimes a_3$ for them, again not denoting but remembering the summation.

A *coalgebra map* is a linear map $f : C \to C'$ which is compatible with the counits e, e' and with the comultiplications; that is, for which the following equalities hold for all $c \in C$

$$e'(f(c)) = e(c), \qquad f(c)_{1'} \otimes f(c)_{2'} = f(c_1) \otimes f(c_2).$$

Proposition 4.3.

(1) For any vector space C over a field k, there is a bijective correspondence between the following structures.

    (i) Opmonoidal structures on the functor $C \otimes - : \mathsf{vec} \to \mathsf{vec}$ (see part 4 of Example 2.5).

    (ii) Coalgebra structures on the vector space C.

(2) For any coalgebras C and C' over k, there is a bijective correspondence between the following data.

    (i) Opmonoidal natural transformations $C \otimes - \to C' \otimes -$ between the opmonoidal functors in part (1).

    (ii) Coalgebra maps $C \to C'$.

Proof.

(1)   If C has a comultiplication $c \mapsto c_1 \otimes c_2$—where Sweedler–Heynemann type implicit summation is understood as in Paragraph 4.2—with counit e, then an opmonoidal structure on the functor $t := C \otimes - : \mathsf{vec} \to \mathsf{vec}$ is provided by the nullary part $e : C \to k$ and the binary part

$$t^2_{V,W} : C \otimes V \otimes W \to C \otimes V \otimes C \otimes W, \qquad c \otimes v \otimes w \mapsto c_1 \otimes v \otimes c_2 \otimes w$$

for any vector spaces V and W. Commutativity of the diagrams of Definition 3.5 is immediate from the coassociativity and counitality of the coalgebra C: the diagrams

$$
\begin{array}{ccc}
c \otimes v \otimes w \otimes z & \xrightarrow{\ t^2_{V \otimes W, Z}\ } & c_1 \otimes v \otimes w \otimes c_2 \otimes z \\[2mm]
\Big\downarrow{\scriptstyle t^2_{V, W \otimes Z}} & & \Big\downarrow{\scriptstyle t^2_{V, W} \otimes 1} \\[2mm]
c_1 \otimes v \otimes c_2 \otimes w \otimes z \xmapsto{\ } c_1 \otimes v \otimes c_{21} \otimes w \otimes c_{22} \otimes z & = & c_{11} \otimes v \otimes c_{12} \otimes w \otimes c_2 \otimes z \\[1mm]
\hspace{3cm}{\scriptstyle 1 \otimes t^2_{W, Z}} & &
\end{array}
$$

$$
\begin{array}{ccc}
c \otimes v & =\!\!=\!\!=\xrightarrow{\ t^2_{k, V}\ } & c_1 \otimes c_2 \otimes v \\[2mm]
\Big\downarrow{\scriptstyle t^2_{V, k}} & & \Big\downarrow{\scriptstyle e \otimes 1 \otimes 1} \\[2mm]
c_1 \otimes v \otimes c_2 \xrightarrow[{\scriptstyle 1 \otimes 1 \otimes e}]{\ } c_1 e(c_2) \otimes v & =\!\!=\!\!= & e(c_1) c_2 \otimes v
\end{array}
$$

commute for any element c of C, for any vector spaces V, W, Z and elements $v \in V$, $w \in W$ and $z \in Z$.

Conversely, an opmonoidal structure (t^0, t^2) on the functor $t := C \otimes -$: vec $\to$ vec determines a coassociative comultiplication $t^2_{k,k} : C \to C \otimes C$ with counit t^0.

Starting with a coalgebra C and iterating these constructions we evidently re-obtain the same coalgebra C. Starting with an opmonoidal functor $(t = C \otimes -, t^0, t^2)$ and iterating these constructions in the opposite order, we clearly re-obtain the nullary part t^0. In order to see that we also re-obtain the same binary part t^2, take any vector spaces V and W; elements $v \in V$ and $w \in W$; and the induced linear maps $v : k \to V$ and $w : k \to W$ sending $\kappa \in k$ to κv and κw, respectively. Then it follows by the naturality of t^2 that the following diagram commutes.

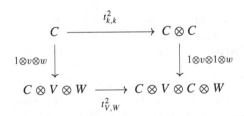

$$
\begin{array}{ccc}
C & \xrightarrow{\ t^2_{k,k}\ } & C \otimes C \\[2mm]
\Big\downarrow{\scriptstyle 1 \otimes v \otimes w} & & \Big\downarrow{\scriptstyle 1 \otimes v \otimes 1 \otimes w} \\[2mm]
C \otimes V \otimes W & \xrightarrow[{\scriptstyle t^2_{V, W}}]{\ } & C \otimes V \otimes C \otimes W
\end{array}
$$

This completes the proof of part (1).

(2) By Exercise 2.8 any natural transformations $C \otimes - \to C' \otimes -$ is induced by a unique linear map $f : C \to C'$; that is, its component at an arbitrary vector space V is $f \otimes 1 : C \otimes V \to C' \otimes V$. In light of part (1), its opmonoidality

translates to the conditions

$$e'(f(c)) = e(c) \quad \text{and} \quad f(c)_{1'} \otimes v \otimes f(c)_{2'} \otimes w = f(c_1) \otimes v \otimes f(c_2) \otimes w$$

for any vector spaces V and W, $v \in V$, $w \in W$ and $c \in C$; where the Sweedler–Heynemann indices 1 and 2 refer to the comultiplication of the coalgebra C and the Sweedler–Heynemann indices $1'$ and $2'$ refer to the comultiplication of the coalgebra C'. This is clearly equivalent to f being a coalgebra map. □

Corollary 4.4. *The category* coalg *whose objects are the coalgebras over a given field k, and whose morphisms are the coalgebra maps, admits the following monoidal structure. The monoidal unit is the base field, regarded as a trivial coalgebra. The monoidal product of any coalgebras C and D is the tensor product vector space $C \otimes D$ with the comultiplication*

$$C \otimes D \to C \otimes D \otimes C \otimes D, \qquad c \otimes d \mapsto c_1 \otimes d_1 \otimes c_2 \otimes d_2$$

—where the Sweedler–Heynemann implicit summation index convention is used for both comultiplications of C and D, see Paragraph 4.2—and the counit $c \otimes d \mapsto e(c)e(d)$—where e denotes both counits of C and D. The monoidal product of some coalgebra homomorphisms h and g is the linear map $h \otimes g$.

Proof. If an opmonoidal functor f : vec → vec is naturally isomorphic to a functor of the form $C \otimes -$ for some vector space C, then by Example 3.6 6 there is a unique opmonoidal structure on $C \otimes -$ which makes it opmonoidally naturally isomorphic to f. So Proposition 4.3 states, in fact, an equivalence between coalg and the category whose objects are those opmonoidal functors which are naturally isomorphic to a functor $C \otimes -$: vec → vec induced by some vector space C; and whose morphisms are the opmonoidal natural transformations.

This latter category has a strict monoidal structure. The monoidal product of the objects is the composition of opmonoidal functors, see Exercise 3.10. The monoidal product of morphisms is the Godement product of opmonoidal natural transformations, see Exercise 3.16. The monoidal unit is the identity functor with the trivial opmonoidal structure in Example 3.6 1 (it is induced by the trivial vector space k).

The equivalence of Proposition 4.3 takes this monoidal structure to the stated monoidal structure of coalg; see Exercise 3.9. □

Definition 4.5. A *bialgebra* over a given field is a vector space T carrying algebra and coalgebra structures such that the multiplication and the unit of the algebra are coalgebra homomorphisms (if regarding the base field and $T \otimes T$ coalgebras as in Corollary 4.4); equivalently, the comultiplication and the counit of the coalgebra are algebra homomorphisms (if regarding the base field and $T \otimes T$ algebras as in Example 3.2 9). Explicitly, the following axioms are required to hold for any elements $a, b \in T$.

$$(ab)_1 \otimes (ab)_2 = a_1 b_1 \otimes a_2 b_2 \quad 1_{T1} \otimes 1_{T2} = 1_T \otimes 1_T \quad e(ab) = e(a)e(b) \quad e(1_T) = 1$$

(where for the comultiplication of the coalgebra T the Sweedler–Heynemann index notation $a \mapsto a_1 \otimes a_2$ is used as in Paragraph 4.2, e stands for the counit of the coalgebra T and 1_T is the unit of the algebra T).

A bialgebra T is a *Hopf algebra* if it admits a map $z : T \to T$—the so-called *antipode*—such that for all $a \in T$, the equalities $a_1 z(a_2) = e(a)1_T = z(a_1)a_2$ hold.

Example 4.6.

1. The linear span of a monoid G is an algebra via the multiplication obtained by the linear extension of the multiplication of G; the unit element of G is a unit for this associative product. On the other hand, the linear map sending each element g of G to $g \otimes g$ is a coassociative comultiplication on the linear span of G with the counit sending each $g \in G$ to the number 1. These algebra and coalgebra structures constitute a bialgebra, known as the *monoid bialgebra*. Clearly, the monoid bialgebra is a Hopf algebra if and only if G is a group; in which case the antipode sends each $g \in G$ to the inverse g^{-1}. This Hopf algebra is called the *group Hopf algebra*. This example is the motivation for general Hopf algebras often being called *quantum groups*.

2. The vector space of functionals on any set G—that is, of the maps from G to the base field with the pointwise linear structure—is an algebra with the pointwise multiplication; the unit is the constant 1 functional. If G is a *finite* monoid then it is a coalgebra as well with comultiplication sending a functional f to the finite sum $\sum_{g \in G} \delta_g \otimes f(g-)$ (where δ_g is the characteristic functional on g; it sends g to 1 and every other element of G to 0). The counit is given by the evaluation on the unit element of G. These algebra and coalgebra structures constitute a bialgebra which is a Hopf algebra if and only if G is a group. In this case the antipode sends any functional f to its composite $f((-)^{-1})$ with the inverse operation.

Exercise 4.7. Show that the antipode of a Hopf algebra T is an algebra homomorphism from T to the opposite algebra T^{op}. Symmetrically, show that the antipode is a coalgebra homomorphism as well from T to the opposite coalgebra.

Theorem 4.8. *For any vector space T (over some field k) there is a bijective correspondence between the following structures.*

(i) *Bimonad structures on the functor $T \otimes - :$ vec $\to$ vec.*
(ii) *Bialgebra structures on the vector space T.*

Moreover, the bimonad in part (i) is a Hopf monad if and only if the bialgebra in part (ii) is a Hopf algebra.

Proof. Proposition 4.1 provides us with a bijection between the monad structures (η, μ) on $t := T \otimes -$ and the algebra structures (η_k, μ_k) on T. Part (1) of Proposition 4.3 provides us with a bijection between the opmonoidal structures (t^0, t^2) on t and the coalgebra structures $(t^0, t^2_{k,k} : c \mapsto c_1 \otimes c_2)$ on T. By part (2) of Proposition 4.3 the multiplication μ and the unit η of the monad t are opmonoidal natural transformations if and only if the multiplication μ_k and the unit η_k of the algebra T are coalgebra homomorphisms. That is, if and only if the algebra and coalgebra structures of T combine to a bialgebra.

Now let T be a bialgebra, equivalently, let $t = T \otimes -$ be a bimonad. Then by Definition 3.20 t is a Hopf monad if and only if the corresponding natural transformation (3.3) is invertible; that is, for any vector spaces V and W, the map

$$\beta_{V,W} : T \otimes V \otimes T \otimes W \to T \otimes V \otimes T \otimes W, \qquad c \otimes v \otimes c' \otimes w \mapsto c_1 \otimes v \otimes c_2 c' \otimes w$$
(4.1)

is invertible. This is clearly equivalent to the invertibility of the map

$$\beta_{k,k} : T \otimes T \to T \otimes T, \qquad c \otimes c' \mapsto c_1 \otimes c_2 c'$$
(4.2)

which is a homomorphism of right modules over the algebra (T, η_k, μ_k) and of left comodules over the coalgebra $(T, t^0, t^2_{k,k})$. In other words, (4.2) renders commutative the following diagrams.

(4.3)

Using the algebra and coalgebra structures of T—but not their compatibility—the hom-tensor adjunction of Example 2.16 and its dual counterpart induce a bijection from the set $\mathsf{mmod}(T \otimes T, T \otimes T)$ of right T-module and left T-comodule maps $T \otimes T \to T \otimes T$ to the set $\mathsf{vec}(T, T)$ of linear maps $T \to T$,

$$b \mapsto T \xrightarrow{1 \otimes \eta_k} T \otimes T \xrightarrow{b} T \otimes T \xrightarrow{t^0 \otimes 1} T$$
(4.4)

with the inverse

$$z \mapsto \quad T \otimes T \xrightarrow{t^2_{k,k} \otimes 1} T \otimes T \otimes T \xrightarrow{1 \otimes z \otimes 1} T \otimes T \otimes T \xrightarrow{1 \otimes \mu_k} T \otimes T. \quad (4.5)$$

The image of $\beta_{k,k} \in \mathsf{mmod}(T \otimes T, T \otimes T)$ of (4.2) under this bijection is the identity map $T \to T$.

The set $\mathsf{mmod}(T \otimes T, T \otimes T)$ carries the structure of a monoid with multiplication provided by the composition of maps and unit provided by the identity map. Via the isomorphism above, this monoid structure can be transferred to $\mathsf{vec}(T, T)$. The resulting multiplication—known as the *convolution product* of the linear maps g and $f : T \to T$—occurs in the right column of the commutative diagram

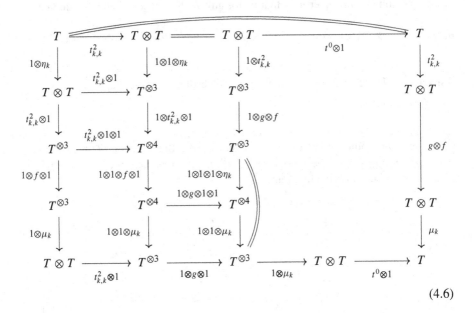

$$(4.6)$$

and the unit is $\eta_k \circ t^0$.

Putting all this together, the map $\beta_{k,k}$ of (4.2) is invertible (with respect to the composition) if and only if the identity map of T has an inverse for the convolution product of (4.6). Such an inverse was called an antipode in Definition 4.5.   □

Since whenever $\beta_{k,k}$ in (4.2) possesses an inverse it is evidently unique, the proof of Theorem 4.8 shows in particular that whenever a bialgebra admits an antipode (i.e. it is a Hopf algebra) then the antipode is unique: equal to

$$T \xrightarrow{1 \otimes \eta_k} T \otimes T \xrightarrow{\beta^{-1}_{k,k}} T \otimes T \xrightarrow{t^0 \otimes 1} T. \quad (4.7)$$

So for a bialgebra being a Hopf algebra is a *property not a further structure*.

4.9. The Category of Modules over a (Hopf) Bialgebra. Consider a bialgebra T over a field k. By Theorem 4.8 it induces a bimonad $T \otimes -$ on **vec**, whose Eilenberg–Moore category is isomorphic to $\mathsf{mod}(T)$ by Proposition 4.1. Hence by Theorem 3.19 there is a monoidal structure on $\mathsf{mod}(T)$ which is lifted from **vec**. Our next aim is to describe it explicitly.

The monoidal unit is the monoidal unit of **vec**; that is, the base field k. By Theorem 3.19 its T-action is provided by the nullary part of the opmonoidal structure of the endofunctor $T \otimes -$ on **vec**. That is, in view of Proposition 4.3, by the counit $T \to k$ of the coalgebra T.

The monoidal product of some T-modules $v : T \otimes V \to V$ and $w : T \otimes W \to W$ lives on the tensor product vector space $V \otimes W$. Again by Theorem 3.19 its T-action is provided by the following composite of the binary part t^2 of the opmonoidal structure of $T \otimes -$ in Proposition 4.3 with the actions v and w,

$$T \otimes V \otimes W \xrightarrow{\ t^2_{V,W}\ } T \otimes V \otimes T \otimes W \xrightarrow{\ v \otimes w\ } V \otimes W$$

$$c \otimes v \otimes w \longmapsto c_1 \otimes v \otimes c_2 \otimes w \longmapsto c_1 \cdot v \otimes c_2 \cdot w$$

(where for the comultiplication of the coalgebra T the Sweedler–Heynemann implicit summation index notation of Paragraph 4.2 is used). This is the so-called *diagonal T-action*.

Assume next that T is a Hopf algebra with some antipode σ; equivalently by Theorem 4.8, $T \otimes -$ is a Hopf monad on **vec**. Then by Theorem 3.27 the left closed structure of **vec** also lifts to $\mathsf{mod}(T)$. That is, the internal hom functor $\mathsf{mod}(T) \to \mathsf{mod}(T)$ is the usual hom functor $\mathsf{vec}(V, -)$ in Example 3.24 2 for any T-module V. Let us compute the T-action on $\mathsf{vec}(V, W)$ for any T-modules V and W (with both actions denoted by $\cdot$).

As we have seen in the proof of Theorem 3.27, for any left T-module V the left adjoint functor $- \otimes V : \mathsf{vec} \to \mathsf{vec}$ lifts to $\mathsf{mod}(T) \to \mathsf{mod}(T)$ along the monad morphism $\beta_{-,V}$ whose component at any vector space X is the morphism

$$T \otimes X \otimes V \xrightarrow{\ t^2_{X,V}\ } T \otimes X \otimes T \otimes V \xrightarrow{\ 1 \otimes 1 \otimes \cdot\ } T \otimes X \otimes V$$

$$c \otimes x \otimes v \longmapsto c_1 \otimes x \otimes c_2 \otimes v \longmapsto c_1 \otimes x \otimes c_2 \cdot v.$$

This has the inverse $c \otimes x \otimes v \mapsto c_1 \otimes x \otimes \sigma(c_2) \cdot v$. So in view of Theorem and Definition 2.33 the internal hom $\mathsf{vec}(V, -)$ of **vec** lifts to $\mathsf{mod}(T)$ along the monad morphism

$$T \otimes \mathsf{vec}(V, W) \xrightarrow{\ \eta^V_{T \otimes \mathsf{vec}(V,W)}\ } \mathsf{vec}(V, T \otimes \mathsf{vec}(V, W) \otimes V) \xrightarrow{\ \mathsf{vec}(V, \beta^{-1}_{\mathsf{vec}(V,W),V})\ }$$

$$c \otimes f \longmapsto c \otimes f \otimes - \longmapsto$$

$$\mathsf{vec}(V, T \otimes \mathsf{vec}(V, W) \otimes V) \xrightarrow{\ \mathsf{vec}(V, 1 \otimes \varepsilon^V_W)\ } \mathsf{vec}(V, T \otimes W)$$

$$c_1 \otimes f \otimes \sigma(c_2) \cdot - \longmapsto c_1 \otimes f(\sigma(c_2) \cdot -)$$

where η^V is the unit and ε^V is the counit of the adjunction $- \otimes V \dashv \mathsf{vec}(V, -) :$ $\mathsf{vec} \to \mathsf{vec}$. This results in the so-called *adjoint T-action* $c \otimes f \mapsto c_1 \cdot f(\sigma(c_2) \cdot -)$ on any linear map $f : V \to W$.

Remark 4.10. Similar steps to those in the proof of Theorem 4.8 prove that the bimonad $T \otimes -$ induced by a bialgebra T on vec is a right Hopf monad in the sense of [30, Section 2.7]—which means that the corresponding natural transformation (3.4) is invertible, see Remark 3.22—if and only if T has a *skew antipode*; that is, a linear map $\widetilde{\sigma} : T \to T$ such that for all $a \in T$, the equalities $\widetilde{\sigma}(a_2)a_1 = e(a)1_T = a_2\widetilde{\sigma}(a_1)$ hold (where 1_T is the unit, e denotes the counit and for the comultiplication the Sweedler–Heynemann implicit summation index convention of Paragraph 4.2 is used).

Consequently, the bimonad $T \otimes -$ on vec is both a left and a right Hopf monad in the sense of [30, Section 2.7] (see again Remark 3.22) if and only if the bialgebra T has both an antipode and a skew antipode. We claim that this is equivalent to the existence of an *invertible* antipode.

Assume first that the bialgebra T has an invertible antipode σ. Then using that by Exercise 4.7 σ is an algebra homomorphism $T \to T^{\mathrm{op}}$, for any $a \in T$

$$\sigma^{-1}(a_2)a_1 = \sigma^{-1}(\sigma(a_1)a_2) = e(a)\sigma^{-1}(1_T) = e(a)1_T$$

and symmetrically, $a_2\sigma^{-1}(a_1) = e(a)1_T$. Thus σ^{-1} is a skew antipode.

Conversely, assume that the bialgebra T has both an antipode σ and a skew antipode $\widetilde{\sigma}$. Since by Exercise 4.7 and its symmetric counterpart both of them are algebra homomorphisms $T \to T^{\mathrm{op}}$, for any $a \in T$

$$\sigma\widetilde{\sigma}(a) = e(a_1)1_T\sigma\widetilde{\sigma}(a_2) = a_1\sigma(a_2)\sigma\widetilde{\sigma}(a_3)$$

$$= a_1\sigma(\widetilde{\sigma}(a_3)a_2) = a_1e(a_2)\sigma(1_T) = a1_T = a$$

and symmetric steps verify $\widetilde{\sigma}\sigma(a) = a$. Thus σ and $\widetilde{\sigma}$ are mutual inverses.

Remark 4.11. Symmetrically to the functor in Theorem 4.8 (i), a vector space T induces another functor $- \otimes T : \textsf{vec} \to \textsf{vec}$. An argument parallel to the proof of Theorem 4.8 leads to a bijection between the bimonad structures on the functor $- \otimes T$ and the bialgebra structures on the vector space T. Furthermore—again in complete analogy with Theorem 4.8 and Remark 4.10—the bimonad $- \otimes T$ induced by a bialgebra T is

- a *right* Hopf monad in the sense of [30, Section 2.7] (see Remark 3.22) if and only if T is a Hopf algebra,
- a Hopf monad in the sense of Definition 5.7 (that is, a *left* Hopf monad in the sense of [30, Section 2.7]; see Remark 3.22) if and only if T has a skew antipode.

Consequently, both conditions hold simultaneously if and only if T is a Hopf algebra with a bijective antipode.

Remark 4.12. The axioms of (Hopf) bialgebra are formally self-dual in the following sense. For a bialgebra T over a field k, let us denote the multiplication and the unit of the constituent algebra by $m : T \otimes T \to T$ and $u : k \to T$, respectively, and let us denote the comultiplication and the counit of the constituent coalgebra by $d : T \to T \otimes T$ and $e : T \to k$, respectively. We can draw the bialgebra axioms as commutative diagrams

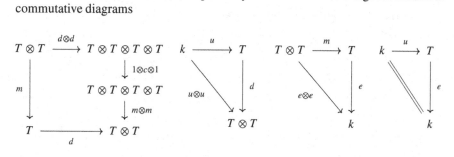

—where $c : T \otimes T \to T \otimes T$ is the flip map $a \otimes a' \mapsto a' \otimes a$—and we can draw the antipode axioms of Hopf algebra in the form of the commutative diagram

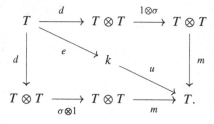

These sets of diagrams are invariant under reversing the arrows and interchanging the roles of the algebra and the coalgebra structures. As an immediate consequence

of this symmetry, the category of comodules over a bialgebra is also monoidal, and the category of comodules over a Hopf algebra is left closed with the structures lifted from vec.

For further reading about Hopf algebras we recommend the books [1, 67, 77, 84, 106, 108].

Chapter 5
(Hopf) Bialgebroids

In this chapter an analysis is carried out which is analogous to, but more general than that in Chap. 4. The category of vector spaces in Chap. 4 is replaced by the category of bimodules over some algebra B; or, isomorphically, the category of left modules over $B \otimes B^{op}$. Those endofunctors on it are considered which are induced, as in Example 2.5 4, by the $B \otimes B^{op}$-module tensor product with a fixed $B \otimes B^{op}$-bimodule A. The monad structures on this functor $A \otimes_{B \otimes B^{op}} -$ are related to the algebra homomorphisms $B \otimes B^{op} \to A$.

The monoidal structure of the category of B-bimodules is explained in Example 3.2 5. The opmonoidal structures with respect to it on the endofunctor $A \otimes_{B \otimes B^{op}} -$, are related to the so-called $B|B$-coring structures on A. This results in a bijection between the bimonads with underlying functor $A \otimes_{B \otimes B^{op}} -$ on the category of B-bimodules; and the *bialgebroids* over the base algebra B. The bijection is shown to restrict to Hopf monads on one hand; and *Hopf algebroids* on the other hand.

The main references for the chapter are [101, 102] and [110].

5.1. Modules, Bimodules and More. For any algebra B over a field, we may consider the *opposite algebra* B^{op}. It lives on the same vector space B but it has the opposite multiplication $b \otimes b' \mapsto b'b$ (where juxtaposition stands for the multiplication of B). Clearly, $(B^{op})^{op}$ is the algebra B.

Consider now the *enveloping algebra* $B^e := B \otimes B^{op}$ (with the factorwise multiplication as in Example 3.2 9). Any B-bimodule V can be regarded as a left module over B^e via the action

$$B^e \otimes V \to V, \qquad b \otimes b' \otimes v \mapsto b \cdot v \cdot b'.$$

Together with the identity map on the morphisms this defines an isomorphism between the category $\mathsf{mod}(B^e)$ of left B^e-modules and the category $\mathsf{bim}(B)$ of B-bimodules.

© Springer Nature Switzerland AG 2018
G. Böhm, *Hopf Algebras and Their Generalizations from a Category Theoretical Point of View*, Lecture Notes in Mathematics 2226,
https://doi.org/10.1007/978-3-319-98137-6_5

Take next a B^e-bimodule A. As in part 4 of Example 2.5, it defines a functor $A \otimes_{B^e} -$: $\mathsf{mod}(B^e) \to \mathsf{mod}(B^e)$. Composing it on both ends with the isomorphism $\mathsf{mod}(B^e) \cong \mathsf{bim}(B)$ of the previous paragraph, we obtain a functor $\mathsf{bim}(B) \to \mathsf{bim}(B)$ which will be denoted by $A \boxtimes -$. Remember that it sends a B-bimodule V to the quotient of the vector space $A \otimes V$ with respect to the subspace

$$\{a \cdot (b \otimes b') \otimes v - a \otimes b \cdot v \cdot b' | a \in A, \ v \in V, \ b \otimes b' \in B^e\}$$

with the B-actions

$$B \otimes (A \boxtimes V) \otimes B \to A \boxtimes V, \qquad b \otimes (a \boxtimes v) \otimes b' \mapsto (b \otimes b') \cdot a \boxtimes v.$$

In particular, considering B^e as a B-bimodule $|B^e$ via the actions

$$B \otimes B^e \otimes B \to B^e, \qquad b \otimes (p \otimes q) \otimes b' \mapsto bp \otimes qb', \tag{5.1}$$

$A \boxtimes |B^e$ is isomorphic to the B-bimodule to be denoted by $|A$; it is A as a vector space and it has the B-actions

$$B \otimes A \otimes B \to A, \qquad b \otimes a \otimes b' \mapsto (b \otimes b') \cdot a. \tag{5.2}$$

Proposition 5.2. *For any algebra B and any B^e-bimodule A, there is a bijective correspondence between the following structures.*

(i) *Monad structures on the functor $A \boxtimes -$: $\mathsf{bim}(B) \to \mathsf{bim}(B)$ (see Paragraph 5.1).*

(ii) *Algebra structures on A together with an algebra homomorphism $B^e \to A$ such that the B^e-actions on A are induced by this homomorphism.*

The structure in part (ii) is called a B^e-ring structure on A.

Furthermore, in this setting the Eilenberg–Moore category of the monad $A \boxtimes -$: $\mathsf{bim}(B) \to \mathsf{bim}(B)$ in part (i) is isomorphic to the category $\mathsf{mod}(A)$ of left modules over the algebra A in part (ii).

Proof. The structure in part (i) consists of a natural transformation η with components $\eta_W : W \to A \boxtimes W$, and a natural transformation μ with components $\mu_W : A \boxtimes (A \boxtimes W) \cong (A \otimes_{B^e} A) \boxtimes W \to A \boxtimes W$, for any B-bimodule W. They are subject to the associativity and unitality conditions.

As proven in Exercise 2.8, the natural transformations η and μ are uniquely determined by their components

$$\eta_{|B^e} : B^e \to A \boxtimes |B^e \cong A \quad \text{and} \quad \mu_{|B^e} : A \otimes_{B^e} A \cong (A \otimes_{B^e} A) \boxtimes |B^e \to A \boxtimes |B^e \cong A$$

which are B^e-bimodule maps. The associativity and unitality conditions on the natural transformations η and μ translate to the commutative diagrams

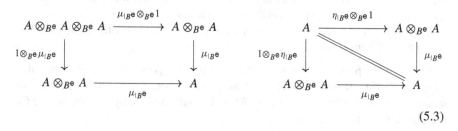

$$(5.3)$$

Given these maps $\eta_{|B^e}$ and $\mu_{|B^e}$, we define a multiplication on A as the composite of the projection $A \otimes A \twoheadrightarrow A \otimes_{B^e} A$ with $\mu_{|B^e} : A \otimes_{B^e} A \to A$. By the associativity of $\mu_{|B^e}$ it is associative and by the unitality of $\mu_{|B^e}$ it possesses a unit given by the image of the unit of B^e under $\eta_{|B^e}$. For this algebra structure $\eta_{|B^e} : B^e \to A$ is unital by construction and it is multiplicative by the commutativity of

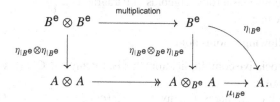

This algebra homomorphism $\eta_{|B^e} : B^e \to A$ induces the given B^e-actions $\cdot$ on A by the commutativity of both diagrams

Conversely, if the data in part (ii) are given, then the associativity of the algebra A implies that the multiplication of A factorizes through $A \otimes_{B^e} A$ via some B^e-bilinear multiplication $A \otimes_{B^e} A \to A$ which is associative in the sense of the first diagram of (5.3). The given algebra homomorphism $B^e \to A$ will be its unit in the sense of the second diagram of (5.3) by the assumption that it induces the B^e-actions on A.

This clearly gives a bijection between the data in parts (i) and (ii).

Concerning the final claim, an Eilenberg–Moore algebra of the monad $A \boxtimes - :$ $\mathrm{bim}(B) \to \mathrm{bim}(B)$ in part (i) is a B-bimodule W together with an associative and unital action $w : A \boxtimes W \to W$. Then W is an A-module via the action provided by the composite of the canonical epimorphism $A \otimes W \twoheadrightarrow A \boxtimes W$ and $w : A \boxtimes W \to W$.

Conversely, if V is a module over the algebra A then the algebra homomorphism $B^e \to A$ induces a B^e-action on V which may be seen as a B-bimodule structure, see Paragraph 5.1. Then the given action $A \otimes V \to V$ clearly factorizes through the epimorphism $A \otimes V \twoheadrightarrow A \boxtimes V$ via the desired action $A \boxtimes V \to V$.

Together with the identity map on the morphisms, these constructions yield the stated mutually inverse functors.   □

Recall from part 5 of Example 3.2 that the category $\mathsf{bim}(B)$ of bimodules over any algebra B is monoidal. Next we wonder about the possible opmonoidal structures on the functor $C \boxtimes -\ : \mathsf{bim}(B) \to \mathsf{bim}(B)$ induced by a B^e-bimodule C; see Paragraph 5.1.

Definition 5.3 ([116, Definition 3.5]). For an arbitrary algebra B, a $B|B$-*coring* consists of

- a B^e-bimodule C
- for the B-bimodule $|C$ of (5.2), a B-bimodule map $\Delta : |C \to |C \otimes_B |C$, $c \mapsto c_1 \otimes_B c_2$ (where implicit summation is understood, analogously to the Sweedler–Heynemann convention for coalgebras in Paragraph 4.2)
- a B-bimodule map $\epsilon : |C \to B$

such that the following axioms hold.

(a) Δ is a coassociative comultiplication; that is, for any $c \in C$, $c_{11} \otimes_B c_{12} \otimes_B c_2 = c_1 \otimes_B c_{21} \otimes_B c_{22}$.
(b) ϵ is the counit of Δ; that is, for any $c \in C$, $(\epsilon(c_1) \otimes 1) \cdot c_2 = c = (1 \otimes \epsilon(c_2)) \cdot c_1$.
(c) Δ respects the right B^e-action as well; in the sense that for any $c \in C$ and $b \otimes b' \in B^e$, $\Delta(c \cdot (b \otimes b')) = c_1 \cdot (b \otimes 1) \otimes c_2 \cdot (1 \otimes b')$.
(d) The counit ϵ satisfies $\epsilon(c \cdot (b \otimes 1)) = \epsilon(c \cdot (1 \otimes b))$ for all $c \in C$ and $b \in B$.

A *morphism of* $B|B$-*corings* is a B^e-bimodule map $f : C \to C'$ such that $\epsilon'(f(c)) = \epsilon(c)$ and $f(c)_{1'} \otimes_B f(c)_{2'} = f(c_1) \otimes_B f(c_2)$ for all $c \in C$.

Exercise 5.4. Show that in any $B|B$-coring C, the image of the comultiplication is central in a suitable B-bimodule; concretely, for any $c \in C$ and $b \in B$, $c_1 \cdot (1 \otimes b) \otimes_B c_2 = c_1 \otimes_B c_2 \cdot (b \otimes 1)$.

Proposition 5.5 ([116, Theorem 3.6 and Corollary 3.8]).

(1) For any algebra B and any B^e-bimodule C, there is a bijective correspondence between the following structures.

  (i) Opmonoidal structures on the functor $C \boxtimes -\ : \mathsf{bim}(B) \to \mathsf{bim}(B)$ (see Paragraph 5.1).
  (ii) $B|B$-coring structures on the B^e-bimodule C.

(2) For $B|B$-corings C and C' there is a bijective correspondence between the following data.

    (i) Opmonoidal natural transformations $C \boxtimes - \;\to\; C' \boxtimes -$ between the opmonoidal functors in part (1).

    (ii) Homomorphisms of $B|B$-corings $C \to C'$.

Proof.

(1) First let a $B|B$-coring structure as in part (ii) be given. The nullary part t^0 of the desired opmonoidal structure in part (i) has the domain

$$C \boxtimes B \cong C \otimes B/\{c \cdot (b \otimes b') \otimes p - c \otimes bpb' | c \in C, \ p \in B, \ b \otimes b' \in B^{\mathrm{e}}\}$$
$$\cong C/\{c \cdot (b \otimes 1) - c \cdot (1 \otimes b) | c \in C, \ b \in B\}.$$

By axiom (d) in Definition 5.3, the counit factorizes through this quotient of C via some B-bimodule map $t^0 : C \boxtimes B \to B$.

The binary part t^2 should have components of the form

$$t^2_{V,W} : C \boxtimes (V \otimes_B W) \to (C \boxtimes V) \otimes_B (C \boxtimes W)$$

for any B-bimodules V and W. We claim that it is meaningful to put

$$t^2_{V,W}(c \boxtimes (v \otimes_B w)) = (c_1 \boxtimes v) \otimes_B (c_2 \boxtimes w); \tag{5.4}$$

all the needed balancing conditions hold. To this end consider the (obviously well-defined) map

$$C \otimes C \otimes V \otimes W \to (C \boxtimes V) \otimes_B (C \boxtimes W), \qquad c \otimes c' \otimes v \otimes w \mapsto (c \boxtimes v) \otimes_B (c' \boxtimes w). \tag{5.5}$$

For any $c, c' \in C, v \in V, w \in W$ and $b \in B$ it satisfies

$$((1 \otimes b) \cdot c \boxtimes v) \otimes_B (c' \boxtimes w) = (c \boxtimes v) \cdot b \otimes_B (c' \boxtimes w)$$
$$= (c \boxtimes v) \otimes_B b \cdot (c' \boxtimes w)$$
$$= (c \boxtimes v) \otimes_B ((b \otimes 1) \cdot c' \boxtimes w).$$

In the first and the last equalities we used the forms of the B-actions on $C \boxtimes V$ and $C \boxtimes W$, respectively, and the second equality holds by the definition of the B-module tensor product. This proves that (5.5) factorizes through the map

$$(C \otimes_B C) \otimes V \otimes W \to (C \boxtimes V) \otimes_B (C \boxtimes W),$$
$$(c \otimes_B c') \otimes v \otimes w \mapsto (c \boxtimes v) \otimes_B (c' \boxtimes w).$$

Pre-composing it with $\Delta \otimes 1 \otimes 1$, we get the map

$$C \otimes V \otimes W \rightarrow (C \boxtimes V) \otimes_B (C \boxtimes W), \qquad c \otimes v \otimes w \mapsto (c_1 \boxtimes v) \otimes_B (c_2 \boxtimes w). \tag{5.6}$$

It satisfies further equalities for any $c \in C$, $v \in V$, $w \in W$ and $b, b' \in B$. First,

$$((c \cdot (b \otimes b'))_1 \boxtimes v) \otimes_B ((c \cdot (b \otimes b'))_2 \boxtimes w) = (c_1 \cdot (b \otimes 1) \boxtimes v) \otimes_B (c_2 \cdot (1 \otimes b') \boxtimes w)$$
$$= (c_1 \boxtimes b \cdot v) \otimes_B (c_2 \boxtimes w \cdot b').$$

The first equality follows by axiom (c) in Definition 5.3 and the second one holds by the definition of the module tensor product $\boxtimes$. Furthermore,

$$(c_1 \boxtimes v \cdot b) \otimes_B (c_2 \boxtimes w) = (c_1 \cdot (1 \otimes b) \boxtimes v) \otimes_B (c_2 \boxtimes w)$$
$$= (c_1 \boxtimes v) \otimes_B (c_2 \cdot (b \otimes 1) \boxtimes w)$$
$$= (c_1 \boxtimes v) \otimes_B (c_2 \boxtimes b \cdot w).$$

The first and the last equalities hold by the definition of the module tensor product $\boxtimes$ and the second one holds by Exercise 5.4. The last two computations prove that (5.6) factorizes through the well-defined map of (5.4). The map of (5.4) is a B-bimodule map since Δ is so:

$$(((b \otimes b') \cdot c)_1 \boxtimes v) \otimes_B (((b \otimes b') \cdot c)_2 \boxtimes w) = ((b \otimes 1) \cdot c_1 \boxtimes v) \otimes_B ((1 \otimes b') \cdot c_2 \boxtimes w)$$
$$= b \cdot (c_1 \boxtimes v) \otimes_B (c_2 \boxtimes w) \cdot b'.$$

The diagrams of Definition 3.5 commute by axioms (a) and (b) in Definition 5.3.

Conversely, let us be given an opmonoidal structure (t^0, t^2) as in part (i). The to-be-counit ϵ is defined as the composite of the B-bimodule epimorphism

$$\mathsf{I}C \rightarrow C/\{c \cdot (b \otimes 1) - c \cdot (1 \otimes b)\} \cong C \boxtimes B$$

with $t^0 : C \boxtimes B \rightarrow B$. By construction it is a B-bimodule map satisfying axiom (d) in Definition 5.3.

The to-be-comultiplication Δ is defined as the composite of the B-bimodule map

$$\mathsf{I}C \rightarrow C \boxtimes (\mathsf{I}B^e \otimes_B \mathsf{I}B^e), \qquad c \mapsto c \boxtimes ((1 \otimes 1) \otimes_B (1 \otimes 1)) \tag{5.7}$$

with $t^2_{\mathsf{I}B^e, \mathsf{I}B^e} : C \boxtimes (\mathsf{I}B^e \otimes_B \mathsf{I}B^e) \rightarrow (C \boxtimes \mathsf{I}B^e) \otimes_B (C \boxtimes \mathsf{I}B^e)$ and with the B-module tensor product of the B-bimodule isomorphisms $C \boxtimes \mathsf{I}B^e \cong \mathsf{I}C$. Then it is a B-bimodule map by construction. Denote the image of any element c of C under this map Δ by $c_1 \otimes_B c_2$ (where implicit summation is understood analogously to the Sweedler–Heynemann convention for coalgebras in Paragraph 4.2).

We turn to checking the validity of the axioms of Definition 5.3. Consider the B-bimodule map

$$- \cdot v \cdot - : {}_I B^e \to V, \qquad b \otimes b' \mapsto b \cdot v \cdot b' \tag{5.8}$$

induced by any element v of an arbitrary B-bimodule V. By the naturality of t^2 the diagram

$$
\begin{array}{ccc}
C \boxtimes ({}_I B^e \otimes_B {}_I B^e) & \xrightarrow{\;t^2_{{}_I B^e, {}_I B^e}\;} & (C \boxtimes {}_I B^e) \otimes_B (C \boxtimes {}_I B^e) \\
{\scriptstyle 1 \boxtimes (- \cdot v \cdot - \otimes_B - \cdot w \cdot -)}\Big\downarrow & & \Big\downarrow {\scriptstyle (1 \boxtimes - \cdot v \cdot -) \otimes_B (1 \boxtimes - \cdot w \cdot -)} \\
C \boxtimes (V \otimes_B W) & \xrightarrow[\;t^2_{V,W}\;]{} & (C \boxtimes V) \otimes_B (C \boxtimes W)
\end{array}
$$

commutes. Its left-bottom path takes $c \boxtimes ((1 \otimes 1) \otimes_B (1 \otimes 1))$ to the expression on the left-hand side of

$$t^2_{V,W}(c \boxtimes (v \otimes_B w)) = (c_1 \boxtimes v) \otimes_B (c_2 \boxtimes w) \tag{5.9}$$

while the top-right path takes $c \boxtimes ((1 \otimes 1) \otimes_B (1 \otimes 1))$ to the expression on the right-hand side, for any elements c of C, v of an arbitrary B-bimodule V and w of an arbitrary B-bimodule W.

With (5.9) at hand, axioms (a) and (b) of Definition 5.3 hold by the commutativity of the diagrams of Definition 3.5; taking each of the occurring objects equal to ${}_I B^e$.

The final axiom (c) follows by the naturality of t^2, as we claim next. Consider the B-bimodule maps of (5.8) induced by the particular elements $b \otimes 1$ and $1 \otimes b$ of the B-bimodule ${}_I B^e$, for an arbitrary fixed element b of B. By the naturality of t^2 the following diagram commutes.

$$
\begin{array}{ccccc}
C \boxtimes ({}_I B^e \otimes_B {}_I B^e) & \xrightarrow{\;t^2_{{}_I B^e, {}_I B^e}\;} & (C \boxtimes {}_I B^e) \otimes_B (C \boxtimes {}_I B^e) & \xrightarrow{\;\cong\;} & C \otimes_B C \\
\Big\downarrow {\scriptstyle 1 \boxtimes (- \cdot (b \otimes 1) \cdot - \otimes_B - \cdot (1 \otimes b') \cdot -)} & & \Big\downarrow {\scriptstyle (1 \boxtimes - \cdot (b \otimes 1) \cdot -) \otimes_B (1 \boxtimes - \cdot (1 \otimes b') \cdot -)} & & \Big\downarrow {\scriptstyle - \cdot (b \otimes 1) \otimes_B - \cdot (1 \otimes b')} \\
C \boxtimes ({}_I B^e \otimes_B {}_I B^e) & \xrightarrow[\;t^2_{{}_I B^e, {}_I B^e}\;]{} & (C \boxtimes {}_I B^e) \otimes_B (C \boxtimes {}_I B^e) & \xrightarrow[\;\cong\;]{} & C \otimes_B C
\end{array}
$$

For any $c \in C$, the value of the left vertical on $c \boxtimes ((1 \otimes 1) \otimes_B (1 \otimes 1))$ is

$$c \boxtimes ((b \otimes 1) \otimes_B (1 \otimes b')) = c \cdot (b \otimes b') \boxtimes ((1 \otimes 1) \otimes_B (1 \otimes 1)).$$

Hence the left-bottom path takes $c \boxtimes ((1 \otimes 1) \otimes_B (1 \otimes 1))$ to $\Delta(c \cdot (b \otimes b'))$; while the top-right path takes it to $c_1 \cdot (b \otimes 1) \otimes c_2 \cdot (1 \otimes b')$. This proves axiom (c).

It remains to see that the above constructions are mutual inverses. Starting with a $B|B$-coring in part (ii) and iterating these constructions we evidently re-obtain the original $B|B$-coring. Starting with an opmonoidal structure (t^0, t^2) in part (i) and iterating the constructions in the opposite order, we obtain an opmonoidal functor whose nullary part is clearly the same map t^0. The binary parts agree by (5.9).

(2) By Exercise 2.8 any natural transformation $C \boxtimes - \rightarrow C' \boxtimes -$ is induced by a unique B^e-bimodule map $f : C \rightarrow C'$; that is, its component at any B-bimodule V is $f \boxtimes 1 : C \boxtimes V \rightarrow C' \boxtimes V$. Opmonoidality of the natural transformation with components $f \boxtimes 1 : C \boxtimes V \rightarrow C' \boxtimes V$ (for any B-bimodule V) is expressed by the commutativity of the diagrams

$$
\begin{array}{ccc}
C \boxtimes B & \xrightarrow{\;t^0\;} & B \\
{\scriptstyle f \boxtimes 1} \downarrow & & \parallel \\
C' \boxtimes B & \xrightarrow[\;t'^0\;]{} & B
\end{array}
\qquad
\begin{array}{ccc}
C \boxtimes (V \otimes_B W) & \xrightarrow{\;t^2_{V,W}\;} & (C \boxtimes V) \otimes_B (C \boxtimes W) \\
{\scriptstyle f \boxtimes 1} \downarrow & & \downarrow {\scriptstyle (f \boxtimes 1) \otimes_B (f \boxtimes 1)} \\
C' \boxtimes (V \otimes_B W) & \xrightarrow[\;t'^2_{V,W}\;]{} & (C' \boxtimes V) \otimes_B (C' \boxtimes W)
\end{array}
$$

for any B-bimodules V and W.

Pre-composing the equal paths around the first one with the canonical epimorphism $C \rightarrow C \boxtimes B$ it gives the equivalent condition $\epsilon' \circ f = \epsilon$.

The top-right path of the second diagram sends an element $c \boxtimes (v \otimes_B w)$ of the domain to $(f(c_1) \boxtimes v) \otimes_B (f(c_2) \boxtimes w)$ while left bottom path sends it to $(f(c)_{1'} \boxtimes v) \otimes_B (f(c)_{2'} \boxtimes w)$. Hence if f satisfies the comultiplicativity condition $f(c)_{1'} \otimes_B f(c)_{2'} = f(c_1) \otimes_B f(c_2)$ then the second diagram commutes. The opposite implication follows by evaluating the diagram at $V = W = {}^|B^e$ and pre-composing its equal paths with the map of (5.7) and post-composing them in both factors with the isomorphism $C \boxtimes {}^|B^e \cong C$. □

Corollary 5.6. *The category of $B|B$-corings admits the following monoidal structure. The monoidal unit is B^e, seen as a B^e-bimodule via the actions provided by the multiplication; with the comultiplication*

$$B^e \rightarrow B^e \otimes_B B^e, \qquad b \otimes b' \mapsto (b \otimes 1) \otimes_B (1 \otimes b')$$

and counit

$$B^e \rightarrow B, \qquad b \otimes b' \mapsto bb'.$$

The monoidal product of some $B|B$-corings C and C' is the B^e-module tensor product $C \otimes_{B^e} C'$ with comultiplication

$$c \otimes_{B^e} c' \mapsto (c_1 \otimes_{B^e} c'_1) \otimes_B (c_2 \otimes_{B^e} c'_2)$$

and counit

$$c \otimes_{B^e} c' \mapsto \epsilon(c \cdot (\epsilon'(c') \otimes 1)) \equiv \epsilon(c \cdot (1 \otimes \epsilon'(c'))).$$

The monoidal product of morphisms is the corresponding B^e-module tensor product.

Proof. This is completely analogous to Corollary 4.4.

If an opmonoidal functor $f : \mathrm{bim}(B) \to \mathrm{bim}(B)$ is naturally isomorphic to a functor of the form $C \boxtimes -$ for some B^e-bimodule C, then by Example 3.6 6 there is a unique opmonoidal structure on $C \boxtimes -$ which makes it opmonoidally naturally isomorphic to f. So Proposition 5.5 states, in fact, an equivalence between the category of $B|B$-corings and the category whose objects are those opmonoidal functors $\mathrm{bim}(B) \to \mathrm{bim}(B)$ which are naturally isomorphic to functors of the form $C \boxtimes -$ for some B^e-bimodule C and whose morphisms are the opmonoidal natural transformations.

This latter category has a strict monoidal structure. The monoidal product of the objects is the composition of opmonoidal functors, see Exercise 3.10. The monoidal product of two objects is again an object in this category by the natural isomorphism $C \boxtimes (C' \boxtimes -) \cong (C \otimes_{B^e} C') \boxtimes -$. The monoidal product of morphisms is the Godement product of opmonoidal natural transformations, see Exercise 3.16. The monoidal unit is the identity functor with the trivial opmonoidal structure in Example 3.6 1. It is an object of the stated category by the natural isomorphism $B^e \boxtimes - \cong 1$.

The equivalence of Proposition 5.5 takes this monoidal structure to the stated monoidal structure of the category of $B|B$-corings; see Exercise 3.9. $\qquad\square$

Definition 5.7. A *bialgebroid* [73, Definition 2.1] over some base algebra B is a vector space equipped with a B^e-ring structure with algebra homomorphism $\eta : B^e \to T$, and a $B|B$-coring structure (Δ, ϵ) on the B^e-bimodule T whose actions are induced by η; such that η and the projected multiplication map $T \otimes_{B^e} T \to T$ are morphisms of $B|B$-corings (where B^e and $T \otimes_{B^e} T$ are understood to have the $B|B$-coring structures in Corollary 5.6). That is, the following identities hold.

(a) For any $c, c' \in T$, $\Delta(cc') = c_1 c'_1 \otimes_B c_2 c'_2$, where a Sweedler type implicit summation index notation as in Paragraph 4.2 is used (note that the right-hand side is meaningful by Exercise 5.4).

(b) For the unit 1_T of the algebra T, $\Delta(1_T) = 1_T \otimes_B 1_T$.

(c) For any $c, c' \in T$, $\epsilon(cc') = \epsilon(c\eta(1_B \otimes \epsilon(c'))) \equiv \epsilon(c\eta(\epsilon(c') \otimes 1_B))$ (the equality of the last two expressions follows by axiom (d) of Definition 5.3).

(d) For the unit 1_T of the algebra T and the unit 1_B of the algebra B, $\epsilon(1_T) = 1_B$.

A *(Schauenburg) Hopf algebroid* over B (or $\times_B$-*Hopf algebra*, by other authors) [102, Theorem & Definition 3.5] is a B-bialgebroid T for which the map from the B^{op}-module tensor product

$$T \otimes_{B^{op}} T := T \otimes T/\{c\eta(1_B \otimes b) \otimes c' - c \otimes \eta(1_B \otimes b)c'|c, c' \in T, \ b \in B\}$$

to the B-module tensor product

$$T \otimes_B T := T \otimes T/\{\eta(1_B \otimes b)c \otimes c' - c \otimes \eta(b \otimes 1_B)c'|c, c' \in C, \ b \in B\}$$

sending $c \otimes_{B^{op}} c'$ to $c_1 \otimes_B c_2 c'$ is invertible (note that it is well-defined by axiom (c) in Definition 5.3).

A slightly different, but equivalent formulation of the bialgebroid axioms occurred much earlier in [115, Definition 4.5] under the name $\times_B$-*bialgebra*; see also the further references [31, Theorem 3.1], [66, pp. 79–80], [101, Definition 4.3] and [121, Definition 3.4] for their comparison.

Remark 5.8. Note that for any algebra B, any B^e-ring $\eta : B^e \to T$ and any left B^e-linear, multiplicative and unital map $\Delta : {}_!T \to {}_!T \otimes_B {}_!T$, the right B^e-linearity axiom (c) of Definition 5.3 identically holds:

$$\Delta(c \cdot (b \otimes b')) = \Delta(c\eta(b \otimes b')) = \Delta(c)\Delta(\eta(b \otimes b')1)$$

$$= c_1\eta(b \otimes 1) \otimes_B c_2\eta(1 \otimes b') = c_1 \cdot (b \otimes 1) \otimes_B c_2 \cdot (1 \otimes b')$$

for all $c \in T$ and $b, b' \in B$.

The following includes information from [101, Theorem 5.1], [102, Theorem & Definition 3.5], [109, Theorem 1.4], [110, Section 4].

Theorem 5.9. *For any algebra B and any B^e-bimodule T, there is a bijective correspondence between the following structures.*

(i) *Bimonad structures on the functor $T \boxtimes - \ : \ \mathsf{bim}(B) \to \mathsf{bim}(B)$ (see Paragraph 5.1).*

(ii) B-*bialgebroid structures on T.*

Furthermore, the bimonad $T \boxtimes - \ : \ \mathsf{bim}(B) \to \mathsf{bim}(B)$ in part (i) is a Hopf monad if and only if the bialgebroid T in part (ii) is a Hopf algebroid over B.

Proof. The bijective correspondence between the data in parts (i) and (ii) follows immediately from Propositions 5.2 and 5.5.

For the bimonad $T \boxtimes - \ : \ \mathsf{bim}(B) \to \mathsf{bim}(B)$ in part (i), the components of the natural transformation (3.3) in Definition 3.20 take the following form, for any B-bimodules V and W, $v \in V$, $w \in W$ and $c, c' \in T$,

$$T \boxtimes (V \otimes_B (T \boxtimes W)) \to (T \boxtimes V) \otimes_B (T \boxtimes W), \tag{5.10}$$

$$c \boxtimes (v \otimes_B (c' \boxtimes w)) \mapsto (c_1 \boxtimes v) \otimes_B (c_2 c' \boxtimes w).$$

This is clearly invertible if the stated map

$$T \otimes_{B^{op}} T \to T \otimes_B T, \qquad c \otimes_{B^{op}} c' \mapsto c_1 \otimes_B c_2 c' \qquad (5.11)$$

is so. Conversely, if (5.10) is invertible for any B-bimodules V and W then it is invertible in particular for $V = W = {}_|B^e$. Composing this isomorphism with the isomorphisms

$$T \boxtimes ({}_|B^e \otimes_B (T \boxtimes {}_|B^e)) \to T \otimes_{B^{op}} T,$$

$$c \boxtimes ((p \otimes p') \otimes_B (c' \boxtimes (b \otimes b'))) \mapsto c\eta(p \otimes 1_B) \otimes_{B^{op}} \eta(p' \otimes 1_B)c'\eta(b \otimes b')$$

and

$$(T \boxtimes {}_|B^e) \otimes_B (T \boxtimes {}_|B^e) \to T \otimes_B T,$$

$$(c \boxtimes (p \otimes p')) \otimes_B (c' \boxtimes (b \otimes b')) \mapsto c\eta(p \otimes p') \otimes_B c\eta(b \otimes b')$$

we obtain the isomorphism (5.11).   □

Let us emphasize that the construction (4.7) of the antipode of a Hopf algebra cannot be generalized to any Hopf algebroid T. The inverse of (5.11) could be applied to $c \otimes 1_B \in T \otimes_B T$, for any $c \in T$, yielding an element of $T \otimes_{B^{op}} T$. However, the counit cannot be applied to the first factor of the result since it is not a right B^{op}-module map in the needed sense. This explains why general Hopf algebroids T do not admit antipodes of the type $T \to T$.

Example 5.10.

1. By Example 3.18 1, the identity functor $\text{bim}(B) \to \text{bim}(B)$ has the canonical structure of an opmonoidal monad. Hence by Example 3.18 2 the same structure is inherited by the naturally isomorphic functor $B^e \boxtimes - : \text{bim}(B) \to \text{bim}(B)$; so by Theorem 5.9 there is a B-bialgebroid structure on B^e. The B^e-ring corresponding to the monad $B^e \boxtimes -$ is provided by the identity algebra homomorphism $B^e \to B^e$; while the $B|B$-coring corresponding to the opmonoidal structure of $B^e \boxtimes -$ was determined in Corollary 5.6.

   Furthermore, the identity functor $\text{bim}(B) \to \text{bim}(B)$ is a Hopf monad by Example 3.21 1. Then by Example 3.21 2 so is the naturally isomorphic functor $B^e \boxtimes - : \text{bim}(B) \to \text{bim}(B)$; so by the application of Theorem 5.9 we infer that the B-bialgebroid B^e of the previous paragraph is a Hopf algebroid.

2. Another example is provided by the so-called *algebraic quantum torus* T_q. Consider a field k and an invertible element q in k.

   As a unital associative algebra, T_q is generated by two invertible elements U and V modulo the relation $VU = qUV$. It has the commutative subalgebra B generated by U, and the algebra homomorphism $B^e \to T_q$, $U^n \otimes U^{n'} \mapsto U^{n+n'}$.

   The comultiplication sends a basis element $U^n V^m$ to the element $U^n V^m \otimes_B V^m$ of ${}_|T_q \otimes_B {}_|T_q$ and the counit sends $U^n V^m$ to $U^n \in B$. This makes T_q a B-bialgebroid.

The map of (5.11), sending $U^n V^m \otimes_{B^{op}} V^{m'}$ to $U^n V^m \otimes_B V^{m'+m}$ has the inverse taking $U^n V^m \otimes_B V^{m'}$ to $U^n V^m \otimes_{B^{op}} V^{m'-m}$. Hence T_q is a Hopf algebroid.

3. For any subalgebra $N \subseteq M$, one can view M as an N-bimodule. Let T be the algebra of endomorphisms of this bimodule. Making certain restrictions on the subalgebra $N \subseteq M$—imposing the so-called *depth 2* and *balancing* conditions—, in [66, Section 4] T was equipped with the structure of a bialgeboid over the subalgebra of M whose elements commute with all elements of N. Under the further assumption that $N \subseteq M$ is a *Frobenius extension*, in [6] T was shown to be a Hopf algebroid.

5.11. The Category of Modules over a (Hopf) Bialgebroid. Consider a bialgebroid T over an arbitrary algebra B; with B^e-ring structure $\eta : B^e \to T$, comultiplication $\Delta : T \to |T \otimes_B |T$, $c \mapsto c_1 \otimes_B c_2$ and counit $\epsilon : |T \to B$. By Theorem 5.9 it induces a bimonad $T \boxtimes -$ on $\mathsf{bim}(B)$, whose Eilenberg–Moore category is isomorphic to $\mathsf{mod}(T)$ by Proposition 5.2. Hence by Theorem 3.19 the monoidal structure of $\mathsf{bim}(B)$ lifts to $\mathsf{mod}(T)$.

The resulting monoidal unit of $\mathsf{mod}(T)$ is the vector space B. By Theorem 3.19 its T-action is provided by the nullary part of the opmonoidal structure of the endofunctor $T \boxtimes -$ on $\mathsf{bim}(B)$. That is, in view of Proposition 5.5, by

$$c \otimes b \mapsto \epsilon(c\eta(b \otimes 1)) = \epsilon(c\eta(1 \otimes b))$$

for all $c \in T$ and $b \in B$. Seen as an algebra homomorphism $c \mapsto \epsilon(c\eta(- \otimes 1)) = \epsilon(c\eta(1 \otimes -))$ from T to the algebra of linear endomorphisms of B, this was called an *anchor map* in [121, Definition 3.4].

The monoidal product of some T-modules

$$T \otimes V \twoheadrightarrow T \boxtimes V \xrightarrow{\overline{v}} V \quad \text{and} \quad T \otimes W \twoheadrightarrow T \boxtimes W \xrightarrow{\overline{w}} W$$

lives on the B-module tensor product vector space $V \otimes_B W$. Again by Theorem 3.19 its T-action is provided by the following composite of the canonical epimorphism, the binary part $t^2_{V,W}$ of the opmonoidal structure of $T \boxtimes -$ in Proposition 5.5 and the projected actions $\overline{v}$ and $\overline{w}$.

$$T \otimes (V \otimes_B W) \twoheadrightarrow T \boxtimes (V \otimes_B W) \xrightarrow{t^2_{V,W}} (T \boxtimes V) \otimes_B (T \boxtimes W) \xrightarrow{\overline{v} \otimes \overline{w}} V \otimes_B W$$

$$c \otimes (v \otimes_B w) \longmapsto c \boxtimes (v \otimes_B w) \longmapsto (c_1 \boxtimes v) \otimes_B (c_2 \boxtimes w) \longmapsto c_1 \cdot v \otimes_B c_2 \cdot w$$

This is again a *diagonal T-action* generalizing that in Paragraph 4.9 for bialgebras.

Assume next that T is a Hopf algebroid; equivalently by Theorem 5.9, $T \boxtimes -$ is a Hopf monad on $\mathsf{bim}(B)$. Then by Theorem 3.27 the left closed structure of $\mathsf{bim}(B)$ in Example 3.24 3 also lifts to $\mathsf{mod}(T)$. This amounts to saying that for any T-modules V and W, the vector space $\mathsf{mod}(B^{op})(V, W)$ of right B-module maps $V \to W$ carries a T-action. In order to write down this action explicitly, for any $c \in T$ we introduce the implicit summation index notation $c^+ \otimes_{B^{op}} c^-$ for the image of $c \otimes_B 1 \in T \otimes_B T$ under the assumed inverse of the map in (5.11). With its help a similar computation to that in Paragraph 4.9 yields the action

$$c \otimes f \mapsto c^+ \cdot f(c^- \cdot -)$$

for any $c \in T$ and any right B-module map $f : V \to W$. This generalizes the *adjoint action* of a Hopf algebra in Paragraph 4.9.

Remark 5.12. Symmetrically to the functor in Theorem 5.9 (i), a B^e-bimodule T induces another functor $- \boxplus T : \mathrm{bim}(B) \to \mathrm{bim}(B)$ where for any B-bimodule V the B^e-module tensor product $V \boxplus T$ is defined as the quotient

$$V \boxplus T := V \otimes T/\{b \cdot v \cdot b' \otimes c - c \otimes (b' \otimes b) \cdot c \mid b, b' \in B, \ v \in V, \ c \in T\}.$$

One can apply an argument parallel to the proof of Theorem 5.9 to relate the bimonad structures on the endofunctor $- \boxplus T$ to some suitable structures on the B^e-bimodule T. In contrast to Remark 4.11 in the case of bimonads on vec and bialgebras, here the answer is not provided by bialgebroids in the same sense of Definition 5.7, but a symmetric variant in [66, pp. 79–80].

The structure that we called a bialgebroid in Definition 5.7 was termed in [66, pp. 79–80] a *left bialgebroid* (by its correspondence to bimonads of the form $T \boxtimes -$ and thus to *left T-modules*). The symmetric structure corresponding to bimonads of the form $- \boxplus T$ was called in [66] a *right bialgebroid*. Summarizing their difference succinctly, the roles of multiplication on the left and on the right are interchanged in their definitions.

5.13. Alternative Notions of Hopf Algebroid. There exist in the literature several other notions also named *Hopf algebroids*.

Lu's Hopf algebroid [73, Definition 4.1] is a B-bialgebroid T such that the canonical epimorphism $T \otimes T \twoheadrightarrow |T \otimes_B| T$ has a linear section i; equipped with an algebra homomorphism—the *Lu antipode*—$\sigma : T \to T^{op}$ subject to the following axioms.

(a) For all $b \in B$, $\sigma \eta(1 \otimes b) = \eta(b \otimes 1)$.

This implies that the map $\sigma \otimes 1 : T \otimes T \to T \otimes T$ projects to

$$\sigma \otimes_B 1 : |T \otimes_B| T \to T \otimes T/\{c\eta(b \otimes 1) \otimes c' - c \otimes \eta(b \otimes 1)c' \mid c, c' \in T, \ b \in B\}.$$

Note that the multiplication $\mu : T \otimes T \to T$ factorizes through the canonical epimorphism

$$T \otimes T \xrightarrow{\quad\quad} T \otimes T/\{c\eta(b \otimes 1) \otimes c' - c \otimes \eta(b \otimes 1)c' \mid c, c' \in T, \ b \in B\}$$

via some projected multiplication $\widehat{\mu} : T \otimes T/\{c\eta(b \otimes 1) \otimes c' - c \otimes \eta(b \otimes 1)c'\} \to T$.

(b) The following diagrams commute.

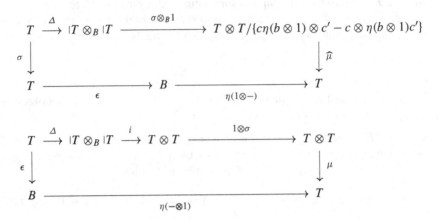

The occurrence of the section i makes this structure conceptually different from a Schauenburg Hopf algebroid in Definition 5.7. The axioms of neither one imply the validity of the axioms of the other.

A *Hopf algebroid in the sense of Böhm and Szlachányi* [13, Definition 4.1] (for a slightly more restrictive definition see the earlier reference [20, Proposition 4.2]) is an algebra T equipped with the structure of a (left) bialgebroid over some base algebra L—with algebra homomorphism $\eta_L : L^e \to T$, comultiplication Δ_L and counit ϵ_L—and also with the structure of right bialgebroid over some base algebra R—with algebra homomorphism $\eta_R : R^e \to T$, comultiplication Δ_R and counit ϵ_R—, together with a linear map (the *antipode*) $\sigma : T \to T$ such that the following axioms hold.

(a) The following diagrams commute.

This makes Δ_L an R-bimodule map and Δ_R an L-bimodule map in a suitable sense so that the following axioms are meaningful.

(b) $(\Delta_L \otimes_R 1) \circ \Delta_R = (1 \otimes_L \Delta_R) \circ \Delta_L$ and $(\Delta_R \otimes_L 1) \circ \Delta_L = (1 \otimes_R \Delta_L) \circ \Delta_R$.

(c) For all $l \in L, r \in R, c \in T, \sigma(\eta_L(1 \otimes l)c\eta_R(1 \otimes r)) = \eta_R(r \otimes 1)\sigma(c)\eta_L(l \otimes 1)$.

This implies that $\sigma \otimes 1 : T \otimes T \to T \otimes T$ projects to

$$\sigma \otimes_L 1 : {}_|T \otimes_L {}_|T \to T \otimes T/\{c\eta_L(l \otimes 1) \otimes c' - c \otimes \eta_L(l \otimes 1)c' | c, c' \in T, \ l \in L\},$$

which can be pre-composed with Δ_L and post-composed with the quotient of the multiplication $\mu_L : T \otimes T/\{c\eta_L(l \otimes 1) \otimes c' - c \otimes \eta_L(l \otimes 1)c'\} \to T$. Symmetrically, $1 \otimes \sigma : T \otimes T \to T \otimes T$ projects to

$$1 \otimes_R \sigma : T_| \otimes_R T_| := T \otimes T/\{c\eta_R(r \otimes 1) \otimes c' - c \otimes c'\eta_R(1 \otimes r) | c, c' \in T, \ r \in R\}$$

$$\to T \otimes T/\{c\eta_R(r \otimes 1) \otimes c' - c \otimes \eta_R(r \otimes 1)c' | c, c' \in T, \ r \in R\}$$

which can be pre-composed with Δ_R and post-composed with the quotient of the multiplication $\mu_R : T \otimes T/\{c\eta_R(r \otimes 1) \otimes c' - c \otimes \eta_R(r \otimes 1)c'\} \to T$.

(d) The following diagrams commute.

From these axioms it follows that the base algebras L and R are in fact anti-isomorphic. In this situation the L-bialgebroid T is a Schauenburg Hopf algebroid as in Definition 5.7, and the right R-bialgebroid T is Hopf algebroid in a symmetric sense, see [13, Section 4.6.2]. The converse is not true, however. Examples of Schauenburg Hopf algebroids with no antipode σ were constructed in [71].

There is an extended literature on bialgebroids and Hopf algebroids, see e.g. [5, 6, 11–13, 15, 17, 20, 21, 31, 40, 41, 64, 66, 101–103, 109–111] and relevant chapters of [99]. Many results on classical bialgebras and Hopf algebras have been extended to them. But there is more than that: they allowed for answers to questions that could not be settled within the classical theory; see e.g. [6] and [66].

Chapter 6
Weak (Hopf) Bialgebras

In this chapter the B-bialgebroids and Hopf algebroids of Chap. 5 are investigated further in the particular case when the base algebra B carries a so-called separable Frobenius structure. Separable Frobenius structures on some algebra B are shown to correspond to separable Frobenius structures on the forgetful functor from the category of B-bimodules to the category of vector spaces. Based on that, the bijection of Chap. 5 between bialgebroids and certain bimonads is refined to bijections between three structures, for any algebra A. First, monoidal structures on the category of A-modules together with separable Frobenius structures on the forgetful functor to the category of vector spaces. Second, bialgebroid structures on A over some base algebra B, together with separable Frobenius structures on B. Finally, *weak bialgebra* structures on A. The B-bialgebroid A is a Hopf algebroid if and only if the corresponding weak bialgebra A is a *weak Hopf algebra*.

The basic references are [22, 104] and [112].

Definition 6.1 ([112, Definition 6.1]). A *separable Frobenius structure* on a functor $f : \mathsf{A}' \to \mathsf{A}$ between monoidal categories consists of

- a monoidal structure $(p^0 : I \to fI', p^2 : f - \otimes f - \to f(- \otimes' -))$
- an opmonoidal structure $(i^0 : fI' \to I, i^2 : f(- \otimes' -) \to f - \otimes f-)$

such that for any objects X, Y and Z of the category A', the equality $p^2_{X,Y} \circ i^2_{X,Y} = 1$ holds and the following diagrams commute.

$$
\begin{array}{ccc}
f(X \otimes' Y) \otimes fZ & \xrightarrow{\; p^2_{X\otimes'Y,Z} \;} & f(X \otimes' Y \otimes' Z) \\[2pt]
{\scriptstyle i^2_{X,Y}\otimes 1} \big\downarrow & & \big\downarrow {\scriptstyle i^2_{X,Y\otimes'Z}} \\[2pt]
fX \otimes fY \otimes fZ & \xrightarrow[\; 1\otimes p^2_{Y,Z} \;]{} & fX \otimes f(Y \otimes' Z)
\end{array}
$$

G. Böhm, *Hopf Algebras and Their Generalizations from a Category Theoretical Point of View*, Lecture Notes in Mathematics 2226,
https://doi.org/10.1007/978-3-319-98137-6_6

$$
\begin{array}{ccc}
fX \otimes f(Y \otimes' Z) & \xrightarrow{\;p^2_{X,Y\otimes'Z}\;} & f(X \otimes' Y \otimes' Z) \\
{\scriptstyle 1\otimes i^2_{Y,Z}}\downarrow & & \downarrow{\scriptstyle i^2_{X\otimes'Y,Z}} \\
fX \otimes fY \otimes fZ & \xrightarrow[\;p^2_{X,Y}\otimes 1\;]{} & f(X \otimes' Y) \otimes fZ
\end{array}
$$

Remark 6.2. For a separable Frobenius functor f from an arbitrary monoidal category A' to vec, and any objects X and Y of A',

$$
\mathsf{Im}(i^2_{X,Y} \circ p^2_{X,Y}) \;\rightarrowtail\; fX \otimes fY \xrightarrow{\;p^2_{X,Y}\;} f(X \otimes' Y)
$$

is a linear isomorphism whose inverse is provided by the corestriction of $i^2_{X,Y}:$ $f(X\otimes' Y) \to fX \otimes fY$. That is to say, the vector space $f(X\otimes' Y)$ can be identified with the range $\mathsf{Im}(i^2_{X,Y}\circ p^2_{X,Y})$ of the idempotent linear map $i^2_{X,Y}\circ p^2_{X,Y}: fX\otimes fY \to fX\otimes fY$.

Example 6.3. Regard a strong monoidal functor f—with nullary part f^0 and binary part f^2—as an opmonoidal functor via the inverses of f^0 and f^2. For these monoidal and opmonoidal structures of f, the diagrams of Definition 6.1 commute by the associativity axiom in Definition 3.5. Since the separability condition $f^2 \circ f^{2-1} = 1$ trivially holds, this gives rise to a separable Frobenius functor.

Exercise 6.4. Show that the composite of separable Frobenius functors is separable Frobenius too, via the monoidal and opmonoidal structures as in Exercise 3.10.

Exercise 6.5. Regard an arbitrary algebra B as a B-bimodule with actions provided by the multiplication, and regard $B \otimes B$ as a B-bimodule with left action provided by the multiplication in the first factor and right action provided by the multiplication in the second factor (that is, the B-bimodule ${}_1B^e$ in (5.1)). Show that any B-bimodule map $B \to B \otimes B$ with respect to these actions is in fact a coassociative (but possibly non-counital) comultiplication.

Definition 6.6. A *separable structure* on an algebra B is a B-bimodule section of the multiplication $m : B \otimes B \to B$ (with respect to the B-actions in Exercise 6.5). That is, a linear map $\delta : B \to B \otimes B$ for which the following diagrams commute.

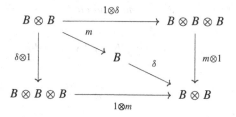
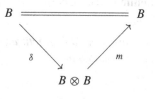

$$(6.1)$$

A *Frobenius structure* on an algebra B is an additional coalgebra structure with comultiplication δ and counit ψ, such that δ is a B-bimodule map with respect to the B-actions in Exercise 6.5; that is, the first diagram of (6.1) commutes.

A *separable Frobenius structure* on an algebra B is a separable structure δ which, if seen as a coassociative comultiplication as in Exercise 6.5, possesses a counit ψ. Equivalently, a separable Frobenius structure on B is a Frobenius structure (ψ, δ) such that δ splits the multiplication m; that is, the second diagram of (6.1) commutes.

Exercise 6.7. In terms of a separable Frobenius structure $(b \mapsto b_{(1)} \otimes b_{(2)}, \psi)$ on an algebra B, construct a separable Frobenius structure on the opposite algebra B^{op}.

Exercise 6.8. Show that any linear map between Frobenius algebras, which is a homomorphism of both algebras and coalgebras, is invertible.

Exercise 6.9. Prove that the following categories are isomorphic (in the sense of Example 2.13 1).

- The category whose objects are the separable Frobenius algebras over a given field; and whose morphisms are those linear maps which are both algebra and coalgebra homomorphisms.
- The category whose objects are the separable Frobenius functors from the monoidal singleton category $\mathbb{1}$ of Example 3.2 7 to the monoidal category **vec** of vector spaces in Example 3.2 2; and whose morphisms are those natural transformations which are both monoidal and opmonoidal.

Lemma 6.10. *Consider a monoidal category* $(\mathsf{A}, J, \square)$ *and a functor* $f : \mathsf{A} \rightarrow$ vec.

(1) Any monoidal structure (p^0, p^2) *on* f *determines an algebra* fJ *with multiplication and unit*

$$fJ \otimes fJ \xrightarrow{\;p^2_{J,J}\;} f(J\square J) \xrightarrow{\;\cong\;} fJ \qquad k \xrightarrow{\;p^0\;} fJ.$$

(2) Any opmonoidal structure (i^0, i^2) *on* f *determines a coalgebra* fJ *with comultiplication and counit*

$$fJ \xrightarrow{\;\cong\;} f(J\square J) \xrightarrow{\;i^2_{J,J}\;} fJ \otimes fJ \qquad fJ \xrightarrow{\;i^0\;} k.$$

(3) For a separable Frobenius structure (p^0, p^2, i^0, i^2) *on* f, *the algebra in part (1) and the coalgebra in part (2) constitute a separable Frobenius algebra.*

Proof. The functor $J : \mathbb{1} \rightarrow \mathsf{A}$ sending the single object of $\mathbb{1}$ to the monoidal unit J is strong monoidal with trivial nullary part $1 : J \rightarrow J$ and binary part $J\square J \cong J$ provided by the (equal left and right) unit constraints. Hence it is a separable Frobenius functor as in Example 6.3.

If f is monoidal, opmonoidal, or separable Frobenius, then the same structure is inherited by the composite functor $\mathbb{1} \xrightarrow{\;J\;} \mathsf{A} \xrightarrow{\;f\;} \mathsf{vec}$ by Exercises 3.10 and 6.4, respectively. Hence it induces the stated algebra, coalgebra or separable Frobenius algebra structure on the vector space fJ by Exercises 3.14 and 6.9, respectively.

□

Proposition 6.11. *For any algebra* B *over a field* k, *there is a bijective correspondence between the following structures.*

(i) Separable Frobenius structures on the monoidal forgetful functor $u : \mathsf{bim}(B)$ *→* vec *of part 3 of Example 3.6.*

(ii) Separable Frobenius structures on the algebra B.

Proof. Assume first that an opmonoidal structure (i^0, i^2) on u is given which—together with the monoidal structure in Example 3.6 3—constitutes a separable Frobenius structure in part (i). Then the structure of a separable Frobenius algebra on uB in part (ii) is obtained from Lemma 6.10 (3).

Conversely, let us be given a separable Frobenius algebra structure on B as in part (ii); then we can proceed as in [112, Lemma 6.4]. The unit element 1 of B is central in the B-bimodule B whose actions are given by multiplication. Since the comultiplication δ of the coalgebra B is a B-bimodule map from this B-bimodule B to ιB^{e} of (5.1), it takes the central element 1 to a central element $1_{(1)} \otimes 1_{(2)} := \delta(1)$ in ιB^{e} (where implicit summation is understood for the Sweedler–Heynemann type indices $\langle 1 \rangle$ and $\langle 2 \rangle$; see Paragraph 4.2). Hence for any B-bimodules V and W there is a well-defined linear map

$$i^2_{V,W} : u(V \otimes_B W) \to uV \otimes uW, \qquad v \otimes_B w \mapsto v \cdot 1_{(1)} \otimes 1_{(2)} \cdot w.$$

In order to see that together with the counit $i^0 : uB \to k$ of the coalgebra B they satisfy the counitality conditions of Definition 3.5 we use the counitality of δ. Namely, for any B-bimodule V,

$$uV \xrightarrow{i^2_{B,V}} uB \otimes uV \xrightarrow{i^0 \otimes 1} uV \qquad uV \xrightarrow{i^2_{V,B}} uV \otimes uB \xrightarrow{1 \otimes i^0} uV$$

$$v \mapsto 1_{(1)} \otimes 1_{(2)} \cdot v \mapsto i^0(1_{(1)})1_{(2)} \cdot v \qquad v \mapsto v \cdot 1_{(1)} \otimes 1_{(2)} \mapsto v \cdot 1_{(1)} i^0(1_{(2)})$$

are identity morphisms. It is immediate to see that the coassociativity condition of Definition 3.5 holds; that the diagrams of Definition 6.1 commute; and that since δ is a section of the multiplication, $p^2 \circ i^2$ is the identity natural transformation. Thus we constructed a separable Frobenius structure on the monoidal functor $u : \mathrm{bim}(B) \to \mathrm{vec}$.

Starting with a separable Frobenius algebra structure in part (ii) and iterating the above constructions, we clearly re-obtain the original counit; and we get the comultiplication $b \mapsto b1_{(1)} \otimes 1_{(2)}$. By its left B-linearity it is equal to the original comultiplication. Starting with the data (i^0, i^2) in part (i) and iterating the above constructions in the opposite order we obtain an opmonoidal structure evidently with the same nullary part i^0. It is more involved to see that we also re-obtain the original binary part i^2. Consider the B-bimodule $|B^e$ of (5.1) and the B-bimodule maps of (5.8). If we denote (understanding implicit summation) by $1_{[1]} \otimes 1_{[2]} \otimes 1_{[3]} \otimes 1_{[4]} \in B \otimes B \otimes B \otimes B$ the value of $i^2_{|B^e,|B^e}$ on $(1 \otimes 1) \otimes_B (1 \otimes 1)$, then the naturality of i^2, more concretely, commutativity of the diagram

$$
\begin{array}{ccc}
|B^e \otimes_B |B^e & \xrightarrow{i^2_{|B^e,|B^e}} & |B^e \otimes |B^e \\
{\scriptstyle (-\cdot v\cdot -) \otimes_B (-\cdot w\cdot -)} \downarrow & & \downarrow {\scriptstyle (-\cdot v\cdot -) \otimes (-\cdot w\cdot -)} \\
V \otimes_B W & \xrightarrow[i^2_{V,W}]{} & V \otimes W
\end{array}
$$

implies that for any $v \otimes_B w \in V \otimes_B W$,

$$i^2_{V,W}(v \otimes_B w) = 1_{[1]} \cdot v \cdot 1_{[2]} \otimes 1_{[3]} \cdot w \cdot 1_{[4]}. \qquad (6.2)$$

In particular, it follows from (6.2) that

$$i^2_{B,B}(1 \otimes_B 1) = 1_{[1]}1_{[2]} \otimes 1_{[3]}1_{[4]}. \qquad (6.3)$$

Evaluate now the first diagram of Definition 6.1 at the B-bimodules $X = Y = \vert B^e$ and an arbitrary B-bimodule Z. Using (6.2), the equal paths around the resulting diagram are seen to take $(1 \otimes 1) \otimes_B (1 \otimes 1) \otimes z$—for any element $z \in Z$—to the equal elements of $\vert B^e \otimes \vert B^e \otimes_B Z$

$$(1_{[1]} \otimes 1_{[2]}) \otimes (1_{[3]} \otimes 1) \otimes_B z \cdot 1_{[4]} = (1_{[1]} \otimes 1_{[2]}) \otimes (1_{[3]} \otimes 1_{[4]}) \otimes_B z$$

from which we infer the equality of elements in $B \otimes B \otimes B \otimes Z$

$$1_{[1]} \otimes 1_{[2]} \otimes 1_{[3]} \otimes z \cdot 1_{[4]} = 1_{[1]} \otimes 1_{[2]} \otimes 1_{[3]} \otimes 1_{[4]} \cdot z. \tag{6.4}$$

Symmetrically, evaluate the second diagram of Definition 6.1 at the B-bimodules $Y = Z = \vert B^e$ and an arbitrary B-bimodule X. Using (6.2) again, the equal paths around the resulting diagram take $x \otimes (1 \otimes 1) \otimes_B (1 \otimes 1)$—for any element $x \in X$—to the equal elements of $X \otimes_B \vert B^e \otimes \vert B^e$

$$1_{[1]} \cdot x \otimes_B (1 \otimes 1_{[2]}) \otimes (1_{[3]} \otimes 1_{[4]}) = x \otimes_B (1_{[1]} \otimes 1_{[2]}) \otimes (1_{[3]} \otimes 1_{[4]})$$

which is equivalent to the equality of elements in $X \otimes B \otimes B \otimes B$

$$1_{[1]} \cdot x \otimes 1_{[2]} \otimes 1_{[3]} \otimes 1_{[4]} = x \cdot 1_{[1]} \otimes 1_{[2]} \otimes 1_{[3]} \otimes 1_{[4]}. \tag{6.5}$$

Combining (6.4) and (6.5) with (6.2) yields $i^2_{V,W}(v \otimes_B w) = v \cdot 1_{[1]} 1_{[2]} \otimes 1_{[3]} 1_{[4]} \cdot w$. Comparing this with (6.3) we have proved that the iteration of our constructions reproduces the original binary part i^2. □

Remark 6.12. Consider a separable Frobenius algebra B. By Proposition 6.11 the forgetful functor $u : \mathsf{bim}(B) \to \mathsf{vec}$ admits a separable Frobenius structure with monoidal part (p^0, p^2) in Example 3.6 3 and opmonoidal structure (i^0, i^2) in the proof of Proposition 6.11. Consequently, by Remark 6.2, the B-module tensor product $V \otimes_B W$ of any B-bimodules V and W can be identified with the image of the idempotent map

$$V \otimes W \xrightarrow{\ \ p^2_{V,W}\ \ } V \otimes_B W \xrightarrow{\ \ i^2_{V,W}\ \ } V \otimes W\ , \qquad v \otimes w \mapsto v \cdot 1_{(1)} \otimes 1_{(2)} \cdot w \tag{6.6}$$

defined in terms of the image $1_{(1)} \otimes 1_{(2)}$ of the unit element of the algebra B under the comultiplication of the coalgebra B (using a Sweedler–Heynemann type implicit summation index notation as in Paragraph 4.2).

Corollary 6.13. *Consider a separable Frobenius algebra B over a field k and a $B\vert B$-coring C. The vector space C is made a coalgebra with the counit obtained as the composite of the counit $C \to B$ of the $B\vert B$-coring C with the counit $B \to k$ of the coalgebra B—that is, the nullary part of the opmonoidal structure of the forgetful functor $\mathsf{bim}(B) \to \mathsf{vec}$ in Proposition 6.11 (i)—and the comultiplication* *obtained as the composite of the comultiplication* $|C \to |C \otimes_B |C$ *of the* $B|B$-*coring C with the component* $|C \otimes_B |C \to C \otimes C$ *of the binary part of the opmonoidal structure of the forgetful functor* $\mathsf{bim}(B) \to \mathsf{vec}$ *in Proposition 6.11 (i).*

Coassociativity and counitality are immediate from the coassociativity and counitality of the $B|B$-*coring C (axioms (a) and (b) of Definition 5.3) and the coassociativity and counitality of the opmonoidal structure of the forgetful functor* $\mathsf{bim}(B) \to \mathsf{vec}$ *(i.e. commutativity of the diagrams of Definition 3.5).*

Proposition 6.14. *For any monoidal category* $(\mathsf{A}, J, \square)$ *there is a bijective correspondence between the following structures.*

(i) Separable Frobenius functors $f : \mathsf{A} \to \mathsf{vec}$.
(ii) • *Separable Frobenius algebras B and*
    • *strict monoidal functors* $F : \mathsf{A} \to \mathsf{bim}(B)$.

Proof. Let us start with a functor $f : \mathsf{A} \to \mathsf{vec}$ equipped with a separable Frobenius structure (P^0, P^2, I^0, I^2) as in part (i). Then the vector space fJ carries the structure of a separable Frobenius algebra in Lemma 6.10.

Observe furthermore that for any object V of A the vector space fV carries an fJ-bimodule structure with action occurring in either path around the commutative diagram

$$
\begin{array}{ccccc}
fJ \otimes fV \otimes fJ & \xrightarrow{\ P^2_{J,V} \otimes 1\ } & f(J\square V) \otimes fJ & \xrightarrow{\ \cong\ } & fV \otimes fJ \\[4pt]
{\scriptstyle 1 \otimes P^2_{V,J}}\Big\downarrow & & & & \Big\downarrow{\scriptstyle P^2_{V,J}} \\[4pt]
fJ \otimes f(V\square J) & & & & f(V\square J) \\[4pt]
{\scriptstyle \cong}\Big\downarrow & & & & \Big\downarrow{\scriptstyle \cong} \\[4pt]
fJ \otimes fV & \xrightarrow[\ \ P^2_{J,V}\ \ ]{} & f(J\square V) & \xrightarrow[\ \cong\ ]{} & fV
\end{array}
\tag{6.7}
$$

and—by the naturality of P^2 and of the unit constraints—f takes any morphism of A to an fJ-bimodule map for this action. This proves that f factorizes through some functor $F : \mathsf{A} \to \mathsf{bim}(fJ)$ via the forgetful functor $u : \mathsf{bim}(fJ) \to \mathsf{vec}$.

It remains to show that F is strict monoidal. Preservation of the monoidal unit amounts to the observation that FJ is the fJ-bimodule living on the vector space $uFJ = fJ$ with fJ-actions given by the multiplication: evaluate the diagram of (6.7) at $V = J$ and recognize in its columns and rows the multiplication on fJ from Lemma 6.10 (1). Concerning the monoidal product of any objects V and W of A, we know from Remark 6.2 that by the separable Frobenius structure of the functor f the vector space $uF(V\square W) = f(V\square W)$ can be identified with the image of the idempotent map

$$fV \otimes fW \xrightarrow{\ P^2_{V,W}\ } f(V\Box W) \xrightarrow{\ I^2_{V,W}\ } fV \otimes fW. \tag{6.8}$$

On the other hand, it was observed in Remark 6.12 that by the separable Frobenius structure of the algebra fJ from Lemma 6.10 (3), the vector space $u(FV \otimes_{fJ} FW)$ can be identified with the image of the idempotent map

$$fV \otimes fW \to fV \otimes fW, \qquad v \otimes w \mapsto v \cdot 1_{(1)} \otimes 1_{(2)} \cdot w \tag{6.9}$$

where $1_{(1)} \otimes 1_{(2)}$ denotes the image of the unit element of the algebra fJ of Lemma 6.10 (1) under the comultiplication of the coalgebra fJ from Lemma 6.10 (2) (with Sweedler–Heynemann type implicit summation understood; see Paragraph 4.2), and the dots stand for the action in (6.7). The equality of the idempotent maps (6.8) and (6.9)—hence the equality of their ranges $uF(V\Box W)$ and $u(FV \otimes_{fJ} FW)$—follows by the commutativity of the following diagram, whose top-right path is the map of (6.9).

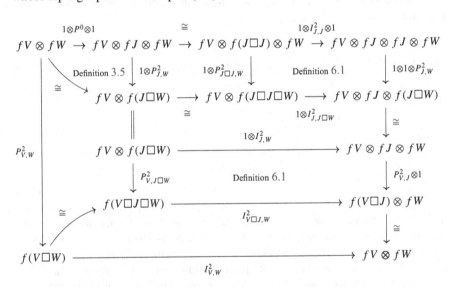

The undecorated regions commute by naturality. Since the left fJ-actions both on $F(V\Box W)$ and $FV \otimes_{fJ} FW$ are given by the left action on FV; and the right fJ-actions both on $F(V\Box W)$ and $FV \otimes_{fJ} FW$ are given by the right action on FW, we conclude the equality of fJ-bimodules $F(V\Box W) = FV \otimes_{fJ} FW$.

Thus we constructed data as in part (ii). Note that with the above identification of the vector space $u(FV \otimes_{fJ} FW) = uF(V\Box W) = f(V\Box W)$ with the image of the equal idempotent linear maps (6.8) and (6.9), the binary part $P^2_{V,W}$ of the monoidal structure of f becomes equal to the corestriction of (6.9), while the binary part $I^2_{V,W}$ of the opmonoidal structure of f becomes equal to the inclusion map.

Conversely, starting with the data in part (ii), the forgetful functor $\mathsf{bim}(B) \to \mathsf{vec}$ carries a separable Frobenius structure by Proposition 6.11. The strict monoidal functor $F : \mathsf{A} \to \mathsf{bim}(B)$ can be seen as a separable Frobenius functor as in Example 6.3. Then their composite has a separable Frobenius structure by Exercise 6.4, hence it provides the datum in part (i).

The bijectivity of this correspondence should be clear from the construction. $\quad\square$

Definition 6.15 ([22, Definition 2.1]). A *weak bialgebra* over a given field is a vector space T equipped with an algebra structure (with unit denoted by 1 and multiplication denoted by juxtaposition) and a coalgebra structure (with comultiplication $\widehat{\Delta} : a \mapsto a_{\hat{1}} \otimes a_{\hat{2}}$—where the Sweedler–Heynemann implicit summation index notation of Paragraph 4.2 is used—and counit $\widehat{\epsilon}$) such that for all $a, a', a'' \in T$ the axioms

(a) $\widehat{\Delta}(aa') = \widehat{\Delta}(a)\widehat{\Delta}(a')$
(b) $(\widehat{\Delta}(1) \otimes 1)(1 \otimes \widehat{\Delta}(1)) = 1_{\hat{1}} \otimes 1_{\hat{2}} \otimes 1_{\hat{3}} = (1 \otimes \widehat{\Delta}(1))(\widehat{\Delta}(1) \otimes 1)$
(c) $\widehat{\epsilon}(aa'_{\hat{1}})\widehat{\epsilon}(a'_{\hat{2}}a'') = \widehat{\epsilon}(aa'a'') = \widehat{\epsilon}(aa'_{\hat{2}})\widehat{\epsilon}(a'_{\hat{1}}a'')$

hold.

The adjective *'weak'* in Definition 6.15 refers to the feature that the comultiplication is not unital; $\widehat{\Delta}(1)$ is not equal to $1 \otimes 1$. This terminology originates from [75].

Clearly, any bialgebra is a weak bialgebra. More interesting examples will be presented in Example 6.25.

Exercise 6.16. Use the same notation as in Definition 6.15. Show that for a weak bialgebra T, the map

$$\epsilon : T \to T, \qquad a \mapsto \widehat{\epsilon}(1_{\hat{1}}a)1_{\hat{2}} \qquad (6.10)$$

satisfies the following identities for any $a, a' \in T$.

(a) $1_{\hat{1}} \otimes \epsilon(1_{\hat{2}}) = 1_{\hat{1}} \otimes 1_{\hat{2}}$ so in particular $\epsilon(1) = 1$
(b) $\widehat{\epsilon}(aa') = \widehat{\epsilon}(a\epsilon(a'))$ so in particular $\widehat{\epsilon}\epsilon(a) = \widehat{\epsilon}(a)$
(c) $\epsilon(aa') = \epsilon(a\epsilon(a'))$ so in particular $\epsilon\epsilon(a) = \epsilon(a)$
(d) $\widehat{\Delta}(a\epsilon(a')) = a_{\hat{1}}\epsilon(a') \otimes a_{\hat{2}}$ and
$\widehat{\Delta}(\epsilon(a')a) = \epsilon(a')a_{\hat{1}} \otimes a_{\hat{2}}$ so in particular $1_{\hat{1}}\epsilon(a) \otimes 1_{\hat{2}} = \epsilon(a)1_{\hat{1}} \otimes 1_{\hat{2}}$
(e) $\epsilon(\epsilon(a)a') = \epsilon(a)\epsilon(a')$
(f) $\epsilon(a) = \epsilon(1_{\hat{1}})\widehat{\epsilon}(1_{\hat{2}}a)$
(g) $\epsilon(a_{\hat{1}}) \otimes a_{\hat{2}} = \epsilon(1_{\hat{1}}) \otimes 1_{\hat{2}}a$ and $a_{\hat{1}} \otimes \epsilon(a_{\hat{2}}) = 1_{\hat{1}}a \otimes 1_{\hat{2}}$
(h) $\epsilon(a)\epsilon(1_{\hat{1}}) \otimes 1_{\hat{2}} = \epsilon(1_{\hat{1}}) \otimes 1_{\hat{2}}\epsilon(a)$
(i) $\epsilon(a_{\hat{1}})a_{\hat{2}} = a$

For a bialgebra the range of the map in (6.10) consists of the scalar multiples of the unit element. For more general weak bialgebras it has a richer structure to be discussed next.

Lemma 6.17. *For any weak bialgebra T over a field k, the range $\epsilon(T)$ of the map in (6.10) carries the structure of a separable Frobenius algebra.*

Proof. Let us use the same notation as in Definition 6.15. By parts (a) and (e) of Exercise 6.16 $\epsilon(T)$ is a unital subalgebra of T. By part (b) of Exercise 6.16 we can define a counit

$$B \to k, \qquad \epsilon(a) \mapsto \widehat{\epsilon}\epsilon(a) = \widehat{\epsilon}(a)$$

and by parts (a) and (h) of Exercise 6.16 we can define a B-bilinear comultiplication

$$B \to B \otimes B, \qquad \epsilon(a) \mapsto \epsilon(a)\epsilon(1_{\widehat{1}}) \otimes 1_{\widehat{2}} = \epsilon(1_{\widehat{1}}) \otimes 1_{\widehat{2}}\epsilon(a).$$

It is a section of the multiplication by part (i) of Exercise 6.16; it is coassociative by Exercise 6.5, and counital by parts (a) and (b) of Exercise 6.16 and since $\widehat{\epsilon}$ is the counit of $\widehat{\Delta}$:

$$\epsilon(a)\epsilon(1_{\widehat{1}}\widehat{\epsilon}(1_{\widehat{2}})) = \epsilon(a)\epsilon(1) \overset{(a)}{=} \epsilon(a) \quad \text{and} \quad \widehat{\epsilon}\epsilon(1_{\widehat{1}})1_{\widehat{2}}\epsilon(a) \overset{(b)}{=} \widehat{\epsilon}(1_{\widehat{1}})1_{\widehat{2}}\epsilon(a) = \epsilon(a).$$

$\square$

Remark 6.18. Observe that the weak bialgebra axioms in Definition 6.15 are invariant both under the replacement of the multiplication with the opposite one, $a \otimes a' \mapsto a'a$; and under the replacement of the comultiplication with the opposite one, $a \mapsto a_{\widehat{2}} \otimes a_{\widehat{1}}$.

Using these invariances, we can introduce symmetric counterparts of the map ϵ of (6.10)

$$\underline{\epsilon}(a) := \widehat{\epsilon}(a1_{\widehat{1}})1_{\widehat{2}} \qquad \overline{\epsilon}(a) := 1_{\widehat{1}}\widehat{\epsilon}(1_{\widehat{2}}a) \qquad \underline{\underline{\epsilon}}(a) := 1_{\widehat{1}}\widehat{\epsilon}(a1_{\widehat{2}}). \qquad (6.11)$$

They satisfy the symmetric variants of the identities in Exercise 6.16.

The following includes information from [104, Theorems 5.1, 5.5 and Corollary 6.2] and [112, Corollary 6.5], see also [91, Theorem 6.1].

Theorem 6.19. *For any algebra T over a field k, there is a bijective correspondence between the following structures.*

(i) • *A monoidal structure on* mod(T) *and*
     • *a separable Frobenius structure on the forgetful functor U :* mod$(T) \to$ vec.

(ii) • *A bialgebroid structure on T (over some unspecified base algebra) and*
     • *a separable Frobenius structure on the base algebra of the bialgebroid.*

(iii) *A weak bialgebra structure on T.*

Proof. Assume first that we are given, as in part (i), a monoidal structure on
$\mathrm{mod}(T)$—with some monoidal product $\square$ and some monoidal unit B—and a
separable Frobenius structure on U. Then a separable Frobenius algebra UB is
obtained as in Lemma 6.10; it should be the base algebra of the desired bialgebroid
in part (ii).

By Proposition 6.14 the separable Frobenius functor $U : \mathrm{mod}(T) \to \mathrm{vec}$
factorizes through a strict monoidal functor $F : \mathrm{mod}(T) \to \mathrm{bim}(UB)$ via the
forgetful functor $u : \mathrm{bim}(UB) \to \mathrm{vec}$. Since both U and u act on the morphisms
as the identity map, so does F. This allows us to view $Fh = h$ as a UB-bimodule
map for any T-module map h.

We aim to interpret $F : \mathrm{mod}(T) \to \mathrm{bim}(UB)$ as the forgetful functor of a
suitable monad (see Paragraph 2.24). By its associativity, any T-action $\triangleright : T \otimes V \to$
V determines a T-module map $T \to V, a \mapsto a \triangleright v$ in terms of any fixed element
v of V. Then it is a UB-bimodule map in the sense that for all $a \in T$ and $b, c \in B$
the equality $(b \cdot a \cdot c) \triangleright v = b \cdot (a \triangleright v) \cdot c$ holds. In particular, choosing V to be the
T-module T with action provided by the multiplication, and choosing v to be the
unit element 1 of the algebra T,

$$(b \cdot 1 \cdot c)(b' \cdot 1 \cdot c') = b \cdot (b' \cdot 1 \cdot c') \cdot c = bb' \cdot 1 \cdot c'c$$

for any elements b, c, b', c' of UB. That is to say,

$$(UB)^{\mathrm{e}} \to T, \qquad b \otimes c \mapsto b \cdot 1 \cdot c$$

is an algebra homomorphism rendering T with the structure of a $(UB)^{\mathrm{e}}$-ring. By
Proposition 5.2 there is a corresponding monad $T \boxtimes - : \mathrm{bim}(UB) \to \mathrm{bim}(UB)$
whose forgetful functor is $F : \mathrm{mod}(T) \to \mathrm{bim}(UB)$.

Since F is its strict monoidal forgetful functor, $T \boxtimes - : \mathrm{bim}(UB) \to \mathrm{bim}(UB)$
is a bimonad by Theorem 3.19. Thus by Theorem 5.9 it determines a UB-
bialgebroid structure on T. Thus we constructed data as in part (ii) from data as
in part (i). Since it was obtained by composing the bijections of Proposition 6.14
and Theorem 5.9, it yields a bijective correspondence.

Assume next that we are given, as in part (ii), a bialgebroid structure on T
over some base algebra B—with B^{e}-ring structure $\eta : B^{\mathrm{e}} \to T$, comultiplication
$\Delta : \lvert T \to \lvert T \otimes_B \lvert T$ and counit $\epsilon : \lvert T \to B$—and a separable Frobenius algebra
structure on B—with comultiplication $b \mapsto b_{\langle 1 \rangle} \otimes b_{\langle 2 \rangle}$ (where the Sweedler–
Heynemann implicit summation index notation of Paragraph 4.2 is applied) and
counit $\psi : B \to k$. For multiple occurrences of the image of the unit element
of B under the comultiplication we will use primed indices $1_{\langle 1 \rangle} \otimes 1_{\langle 2 \rangle}$, $1_{\langle 1' \rangle} \otimes 1_{\langle 2' \rangle}$,
$1_{\langle 1'' \rangle} \otimes 1_{\langle 2'' \rangle}$ and so on.

A coalgebra structure $(\widehat{\Delta}, \widehat{\epsilon})$ on T is obtained from Corollary 6.13. We are facing
a situation where two comultiplications are present: Δ for the $B\lvert B$-coring T and
$\widehat{\Delta}$ for the coalgebra T. We use two variants of Sweedler–Heynemann's implicit
summation index notation in Paragraph 4.2 for them. We write $\Delta(a) = a_1 \otimes_B a_2$

and $\widehat{\Delta}(a) = a_{\hat{1}} \otimes a_{\hat{2}}$ for any $a \in T$. They are related by $a_{\hat{1}} \otimes a_{\hat{2}} := \eta(1 \otimes 1_{\langle 1 \rangle})a_1 \otimes \eta(1_{\langle 2 \rangle} \otimes 1)a_2$; and they have the respective counits ϵ and $\widehat{\epsilon} := \psi \circ \epsilon$.

Let us turn to checking the compatibility axioms between the algebra and coalgebra structures of T. Axiom (a) in Definition 6.15 holds for any $a, a' \in T$ by

$$\widehat{\Delta}(a)\widehat{\Delta}(a') = \eta(1 \otimes 1_{\langle 1 \rangle})a_1\eta(1 \otimes 1_{\langle 1' \rangle})a_1' \otimes \eta(1_{\langle 2 \rangle} \otimes 1)a_2\eta(1_{\langle 2' \rangle} \otimes 1)a_2'$$
$$= \eta(1 \otimes 1_{\langle 1 \rangle})a_1a_1' \otimes \eta(1_{\langle 2 \rangle} \otimes 1)a_2\eta(1_{\langle 1' \rangle} \otimes 1)\eta(1_{\langle 2' \rangle} \otimes 1)a_2'$$
$$= \eta(1 \otimes 1_{\langle 1 \rangle})a_1a_1' \otimes \eta(1_{\langle 2 \rangle} \otimes 1)a_2a_2'$$
$$= \eta(1 \otimes 1_{\langle 1 \rangle})(aa')_1 \otimes \eta(1_{\langle 2 \rangle} \otimes 1)(aa')_2 = \widehat{\Delta}(aa').$$

The second equality follows by Exercise 5.4; in the third equality we used that

$$\eta(1_{\langle 1' \rangle} \otimes 1)\eta(1_{\langle 2' \rangle} \otimes 1) = \eta(1_{\langle 1' \rangle}1_{\langle 2' \rangle} \otimes 1) = \eta(1 \otimes 1) = 1 \otimes 1$$

by the fact that the comultiplication of B is a section of the multiplication; and in the penultimate equality we used the multiplicativity of Δ; that is, axiom (a) in Definition 5.7.

The comultiplication Δ of the bialgebroid T is unital by axiom (b) Definition 5.7 from which it follows that

$$\widehat{\Delta}(1) = \eta(1 \otimes 1_{\langle 1 \rangle}) \otimes \eta(1_{\langle 2 \rangle} \otimes 1).$$

Then

$$(\widehat{\Delta}(1) \otimes 1)(1 \otimes \widehat{\Delta}(1)) = \eta(1 \otimes 1_{\langle 1 \rangle}) \otimes \eta(1_{\langle 2 \rangle} \otimes 1)\eta(1 \otimes 1_{\langle 1' \rangle}) \otimes \eta(1_{\langle 2' \rangle} \otimes 1)$$
$$= \eta(1 \otimes 1_{\langle 1 \rangle}) \otimes \eta(1_{\langle 2 \rangle} \otimes 1_{\langle 1' \rangle}) \otimes \eta(1_{\langle 2' \rangle} \otimes 1)$$
$$= \eta(1 \otimes 1_{\langle 1 \rangle}) \otimes \eta(1 \otimes 1_{\langle 1' \rangle})\eta(1_{\langle 2 \rangle} \otimes 1) \otimes \eta(1_{\langle 2' \rangle} \otimes 1)$$
$$= (1 \otimes \widehat{\Delta}(1))(\widehat{\Delta}(1) \otimes 1)$$

and since Δ is a B-bimodule map this is also equal to $1_{\hat{1}} \otimes 1_{\hat{2}} \otimes 1_{\hat{3}}$, proving that axiom (b) in Definition 6.15 holds.

Finally, since Δ is a right B^e-module map in the sense of axiom (c) in Definition 5.3, it follows that

$$\widehat{\Delta}(a\eta(b \otimes b')) = i^2_{|T,|T}(\Delta(a\eta(b \otimes b')))$$
$$= i^2_{|T,|T}(a_1\eta(b \otimes 1) \otimes a_2\eta(1 \otimes b')) = a_{\hat{1}}\eta(b \otimes 1) \otimes a_{\hat{2}}\eta(1 \otimes b')$$

for any $a \in T$ and $b, b' \in B$. On the other hand, by axiom (c) in Definition 5.7,

$$\widehat{\epsilon}(aa') = \psi\epsilon(aa') = \psi\epsilon(a\eta(1 \otimes \epsilon(a'))) = \widehat{\epsilon}(a\eta(1 \otimes \epsilon(a'))) \qquad \text{and}$$

$$\widehat{\epsilon}(aa') = \psi\epsilon(aa') = \psi\epsilon(a\eta(\epsilon(a') \otimes 1)) = \widehat{\epsilon}(a\eta(\epsilon(a') \otimes 1))$$

for all $a, a' \in T$. Using these identities together with the fact that $\widehat{\epsilon}$ is the counit of $\widehat{\Delta}$, it follows for any $a, a', a'' \in T$ that

$$\widehat{\epsilon}(aa'_{\widehat{1}})\widehat{\epsilon}(a'_{\widehat{2}}a'') = \widehat{\epsilon}(aa'_{\widehat{1}})\widehat{\epsilon}(a'_{\widehat{2}}\eta(1 \otimes \epsilon(a'')))$$

$$= \widehat{\epsilon}(a[a'\eta(1 \otimes \epsilon(a''))]_{\widehat{1}}\widehat{\epsilon}([a'\eta(1 \otimes \epsilon(a''))]_{\widehat{2}}))$$

$$= \widehat{\epsilon}(aa'\eta(1 \otimes \epsilon(a''))) = \widehat{\epsilon}(aa'a'') \qquad \text{and}$$

$$\widehat{\epsilon}(aa'_{\widehat{2}})\widehat{\epsilon}(a'_{\widehat{1}}a'') = \widehat{\epsilon}(aa'_{\widehat{2}})\widehat{\epsilon}(a'_{\widehat{1}}\eta(\epsilon(a'') \otimes 1))$$

$$= \widehat{\epsilon}(a[a'\eta(\epsilon(a'') \otimes 1)]_{\widehat{2}}\widehat{\epsilon}([a'\eta(\epsilon(a'') \otimes 1)]_{\widehat{1}}))$$

$$= \widehat{\epsilon}(aa'\eta(\epsilon(a'') \otimes 1)) = \widehat{\epsilon}(aa'a'')$$

so that axiom (c) in Definition 6.15 also holds. This completes the construction of the data in part (iii) from those in part (ii).

Conversely, assume that a coalgebra structure $(\widehat{\Delta}, \widehat{\epsilon})$ on T as in part (iii) is given. By Lemma 6.17, the range $\epsilon(T)$ of the map (6.10) is a separable Frobenius algebra. It should be the base algebra of the bialgebroid in part (ii) to be constructed next.

Recall from (6.11) the idempotent map $\bar{\epsilon}$ whose image is another unital subalgebra of T. By part (b) of Exercise 6.16, $\bar{\epsilon} \circ \epsilon = \bar{\epsilon}$ and symmetrically, $\epsilon \circ \bar{\epsilon} = \epsilon$. Hence they restrict to mutually inverse isomorphisms between the vector spaces $\epsilon(T)$ and $\bar{\epsilon}(T)$. Using Axiom (c) in Definition 6.15 in the second equality, part (d) of Exercise 6.16 in the third one, and its part (b) in the fourth one,

$$\bar{\epsilon}(\epsilon(a)a') = 1_{\widehat{1}}\widehat{\epsilon}(1_{\widehat{2}}\epsilon(a)a') = 1_{\widehat{1}}\widehat{\epsilon}(1_{\widehat{2}}\epsilon(a)_{\widehat{1}})\widehat{\epsilon}(\epsilon(a)_{\widehat{2}}a') \qquad (6.12)$$

$$= 1_{\widehat{1}}\widehat{\epsilon}(1_{\widehat{2}}1_{\widehat{1}'}\epsilon(a))\widehat{\epsilon}(1_{\widehat{2}'}a') = 1_{\widehat{1}}\widehat{\epsilon}(1_{\widehat{2}}1_{\widehat{1}'}a)\widehat{\epsilon}(1_{\widehat{2}'}a') = \bar{\epsilon}(\bar{\epsilon}(a')a)$$

for any $a, a' \in T$. Symmetrically to part (e) of Exercise 6.16, $\bar{\epsilon}(\bar{\epsilon}(a')a) = \bar{\epsilon}(a')\bar{\epsilon}(a)$ and using this identity we infer from (6.12) that

$$\bar{\epsilon}(\epsilon(a)\epsilon(a')) = \bar{\epsilon}(\bar{\epsilon}\epsilon(a')a) = \bar{\epsilon}\epsilon(a')\bar{\epsilon}(a) = \bar{\epsilon}\epsilon(a')\bar{\epsilon}\epsilon(a)$$

for any $a, a' \in T$. Since ϵ is unital by part (a) of Exercise 6.16 and $\bar{\epsilon}$ is so by symmetry, this proves that the subalgebras $\epsilon(T)$ and $\bar{\epsilon}(T)$ of T are anti-isomorphic. Moreover, the elements of $\bar{\epsilon}(T)$ commute with the elements of $\epsilon(T)$ by part (d) of Exercise 6.16. Hence there is an algebra homomorphism

$$\epsilon(T) \otimes \epsilon(T)^{\text{op}} \to T, \qquad \epsilon(a) \otimes \epsilon(a') \mapsto \epsilon(a)\bar{\epsilon}(a') = \bar{\epsilon}(a')\epsilon(a)$$

rendering T with the structure of a $\epsilon(T)^{\text{e}}$-ring.

The candidate comultiplication of the $\epsilon(T)$-bialgebroid T is the composite map

$$\Delta: \quad T \xrightarrow{\ \widehat{\Delta}\ } T \otimes T \longrightarrow |T \otimes_{\epsilon(T)} |T \ .$$

By the second identity in part (d) of Exercise 6.16 and its symmetric counterpart

$$\widehat{\Delta}(\bar{\epsilon}(a)a') = a'_{\hat{1}} \otimes \bar{\epsilon}(a)a'_{\hat{2}}, \tag{6.13}$$

for all $a, a' \in T$, Δ is an $\epsilon(T)$-bimodule map $|T \rightarrow |T \otimes_{\epsilon(T)} |T$. The candidate counit is the corestriction of (6.10) to a map $\epsilon : T \rightarrow \epsilon(T)$. Symmetrically to (6.12), $\epsilon(\bar{\epsilon}(a')a) = \epsilon(\epsilon(a)a')$ for all $a, a' \in T$. Together with part (e) of Exercise 6.16 this proves $\epsilon(\epsilon(a')\bar{\epsilon}(a'')a) = \epsilon(a')\epsilon(a)\epsilon(a'')$ for all $a, a', a'' \in T$; that is, the required bilinearity of $\epsilon : |T \rightarrow \epsilon(T)$. The comultiplication Δ is coassociative by the coassociativity of $\widehat{\Delta}$ and counital by part (i) of Exercise 6.16 and its symmetric counterpart

$$\bar{\epsilon}(a_{\hat{2}})a_{\hat{1}} = a \tag{6.14}$$

for all $a \in T$. By the multiplicativity of $\widehat{\Delta}$, Δ is also multiplicative and it is unital by

$$1_{\hat{1}} \otimes_B 1_{\hat{2}} = 1_{\hat{1}} \otimes_B \epsilon(1_{\hat{2}})1 = \bar{\epsilon}(1_{\hat{2}})1_{\hat{1}} \otimes_B 1 = 1 \otimes_B 1.$$

Here the first equality follows by part (a) of Exercise 6.16, the second equality holds by definition of the occurring B-module tensor product and the last equality is a consequence of (6.14). That is, axioms (a) and (b) in Definition 5.7 hold. Then from its left B^e-linearity the right B^e-linearity of Δ in the sense of axiom (c) in Definition 5.3 also follows, see Remark 5.8. The counit ϵ preserves the unit by part (a) of Exercise 6.16 so that axiom (d) in Definition 5.7 holds. Finally, axiom (d) in Definition 5.3 and axiom (c) in Definition 5.7 follow by part (c) of Exercise 6.16 and its symmetric counterpart $\epsilon(aa') = \epsilon(a\bar{\epsilon}(a'))$ for all $a, a' \in T$. Summarizing, we have constructed a B-bialgebroid T completing the construction of the data in part (iii) from those in part (ii). Note that in this situation the equality

$$1_{\hat{1}}\bar{\epsilon}(a) \otimes 1_{\hat{2}} = 1_{\hat{1}} \otimes 1_{\hat{2}}\epsilon(a) \tag{6.15}$$

holds for any $a \in T$ by Exercise 5.4 (or by applying $\bar{\epsilon} \otimes 1$ to the identity of part (h) of Exercise 6.16).

For any B-bialgebroid T, the map

$$B \rightarrow T, \qquad b \mapsto \eta(b \otimes 1) \tag{6.16}$$

is a monomorphism split by the counit ϵ. Hence we may identify the base algebra B with its isomorphic image in T under (6.16). Up to this identification, the above constructions between the data in parts (ii) and (iii) are clearly mutual inverses. □

By the *base algebra* of a weak bialgebra we mean the base algebra of the corresponding bialgebroid in Theorem 6.19 (ii); that is, the range of the map (6.10).

Definition 6.20. A *weak Hopf algebra* is a weak bialgebra T together with a linear map $\sigma : T \to T$—the so-called *antipode*—such that for all $a \in T$ the following axioms hold

$$a_{\hat{1}}\sigma(a_{\hat{2}}) = \widehat{\epsilon}(1_{\hat{1}}a)1_{\hat{2}}, \qquad \sigma(a_{\hat{1}})a_{\hat{2}} = 1_{\hat{1}}\widehat{\epsilon}(a1_{\hat{2}}), \qquad \sigma(a_{\hat{1}})a_{\hat{2}}\sigma(a_{\hat{3}}) = \sigma(a).$$

Here—as in Definition 6.15—multiplication is denoted by juxtaposition, 1 denotes the unit of the algebra T, for the comultiplication $a \mapsto a_{\hat{1}} \otimes a_{\hat{2}}$ the Sweedler–Heynemann implicit summation index notation of Paragraph 4.2 is used and $\widehat{\epsilon}$ denotes the counit of the coalgebra T.

Clearly, any Hopf algebra is a weak Hopf algebra. More interesting examples will be presented in Example 6.25.

Exercise 6.21. Prove that the antipode of a weak Hopf algebra T is an algebra homomorphism from T to its opposite T^{op}.

An abstract argument for the antipode of a weak Hopf algebra being a coalgebra homomorphism too will be given in Remark 6.27.

Theorem 6.22. *A weak bialgebra is a weak Hopf algebra if and only if the corresponding bialgebroid in part (ii) of Theorem 6.19 is a Hopf algebroid.*

Proof. Since a weak bialgebra T carries both algebra and coalgebra structures (with structural maps denoted as in Definition 6.20), we have the mutually inverse isomorphisms (4.4) and (4.5) between the convolution algebra $\mathsf{vec}(T, T)$ of linear maps $T \to T$—whose multiplication in the right column of (4.6) will be denoted by $*$—and the algebra $\mathsf{mmod}(T \otimes T, T \otimes T)$ of left T-comodule right T-module maps $T \otimes T \to T \otimes T$—with multiplication provided by the composition of maps.

The antipode axioms of a weak Hopf algebra in Definition 6.20 can be interpreted as identities in the convolution algebra $\mathsf{vec}(T, T)$

$$1 * \sigma = \epsilon, \qquad \sigma * 1 = \overline{\epsilon}, \qquad \sigma * 1 * \sigma = \sigma. \tag{6.17}$$

The isomorphism (4.5) takes the identity map $T \to T$ to

$$\beta : T \otimes T \to T \otimes T, \qquad a \otimes a' \mapsto a_{\hat{1}} \otimes a_{\hat{2}}a' \tag{6.18}$$

and it takes ϵ and $\bar{\epsilon}$ to the respective idempotent maps

$$\Xi : T \otimes T \to T \otimes T, \qquad a \otimes a' \mapsto a_{\hat{1}} \otimes \epsilon(a_{\hat{2}})a' = 1_{\hat{1}}a \otimes 1_{\hat{2}}a' \qquad \text{and}$$

$$\underline{\Xi} : T \otimes T \to T \otimes T, \qquad a \otimes a' \mapsto a_{\hat{1}} \otimes \underline{\bar{\epsilon}}(a_{\hat{2}})a' = a1_{\hat{1}} \otimes \underline{\bar{\epsilon}}(1_{\hat{2}})a', \quad (6.19)$$

where the last equality of the first line follows by the second identity of part (g) of Exercise 6.16, and the last equality of the second line follows by a symmetric variant the first identity of part (g) of Exercise 6.16. So the existence of a linear map $\sigma : T \to T$ satisfying the antipode axioms (6.17) is equivalent to the existence of a left T-comodule right T-module map $\tilde{\beta} : T \otimes T \to T \otimes T$ such that

$$\beta \circ \tilde{\beta} = \Xi, \qquad \tilde{\beta} \circ \beta = \underline{\Xi}, \qquad \tilde{\beta} \circ \beta \circ \tilde{\beta} = \tilde{\beta}. \qquad (6.20)$$

On the other hand, the $\epsilon(T)$-bialgebroid T in Theorem 6.19 (ii) is a Hopf algebroid if (5.11); that is, the following map is invertible

$$T \otimes_{\epsilon(T)^{\mathrm{op}}} T \to T \otimes_{\epsilon(T)} T, \qquad a \otimes_{\epsilon(T)^{\mathrm{op}}} a' \mapsto a_{\hat{1}} \otimes_{\epsilon(T)} a_{\hat{2}}a'. \qquad (6.21)$$

By Remark 6.12 its domain is isomorphic to the range of the idempotent map

$$T \otimes T \to T \otimes T, \qquad a \otimes a' \mapsto a\bar{\epsilon}(1_{\hat{2}}) \otimes \bar{\epsilon}\epsilon(1_{\hat{1}})a' = a\bar{\epsilon}(1_{\hat{2}}) \otimes 1_{\hat{1}}a' = a1_{\hat{1}} \otimes \underline{\bar{\epsilon}}(1_{\hat{2}})a',$$

where the first equality holds since by part (b) of Exercise 6.16 $\bar{\epsilon} \circ \epsilon = \bar{\epsilon}$; and by a symmetric variant of part (a) of Exercise 6.16 $\bar{\epsilon}(1_{\hat{1}}) \otimes 1_{\hat{2}} = 1_{\hat{1}} \otimes 1_{\hat{2}}$, and the last equality follows by the explicit forms of the maps $\underline{\epsilon}$ and $\bar{\epsilon}$ in (6.11). Again by Remark 6.12 and the same identities, the codomain of (6.21) is isomorphic to the range of the idempotent map

$$T \otimes T \to T \otimes T, \qquad a \otimes a' \mapsto \bar{\epsilon}\epsilon(1_{\hat{1}})a \otimes 1_{\hat{2}}a' = 1_{\hat{1}}a \otimes 1_{\hat{2}}a'.$$

Composing (6.21) with these isomorphisms, we conclude that the $\epsilon(T)$-bialgebroid T in Theorem 6.19 (ii) is a Hopf algebroid if and only if

$$\omega : T1_{\hat{1}} \otimes \underline{\bar{\epsilon}}(1_{\hat{2}})T \to 1_{\hat{1}}T \otimes 1_{\hat{2}}T, \qquad (6.22)$$

$$a1_{\hat{1}} \otimes \underline{\bar{\epsilon}}(1_{\hat{2}})a' \mapsto (a1_{\hat{1}})_{\hat{1}} \otimes (a1_{\hat{1}})_{\hat{2}}\underline{\bar{\epsilon}}(1_{\hat{2}})a' = a_{\hat{1}} \otimes a_{\hat{2}}1_{\hat{1}}\underline{\bar{\epsilon}}(1_{\hat{2}})a' = a_{\hat{1}} \otimes a_{\hat{2}}a'$$

is invertible. (The equalities follow by symmetric variants of parts (d) and (i) of Exercise 6.16, respectively.)

If ω of (6.22) is invertible then we construct $\tilde{\beta}$ as the composite

$$T \otimes T \xrightarrow{\; 1_{\hat{1}} - \otimes 1_{\hat{2}} - \;} 1_{\hat{1}}T \otimes 1_{\hat{2}}T \xrightarrow{\;\omega^{-1}\;} T1_{\hat{1}} \otimes \underline{\bar{\epsilon}}(1_{\hat{2}})T \rightarrowtail T \otimes T$$

of left T-comodule right T-module maps. By the commutativity of both diagrams

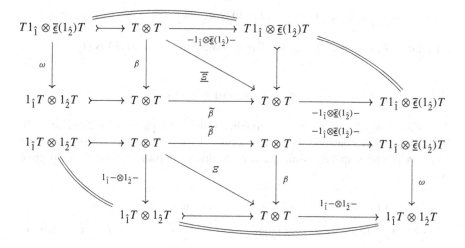

it satisfies the first two equalities of (6.20). The third equality of (6.20) follows from them since $\tilde{\beta} \circ \Xi = \tilde{\beta} = \overline{\Xi} \circ \tilde{\beta}$ is evident from the construction.

Conversely, if there exists a map $\tilde{\beta}$ satisfying the equalities of (6.20) then

$$1_{\hat{1}}T \otimes 1_{\hat{2}}T \;\rightarrowtail\; T \otimes T \xrightarrow{\;\tilde{\beta}\;} T \otimes T \xrightarrow{\;-1_{\hat{1}}\otimes\underline{\bar{\varepsilon}}(1_{\hat{2}})-\;} T1_{\hat{1}} \otimes \underline{\bar{\varepsilon}}(1_{\hat{2}})T \qquad (6.23)$$

is the inverse of ω by the commutativity of the following diagrams.

it satisfies the first two equalities...

6.23. **Uniqueness of the Antipode.** The proof of Theorem 6.22 shows that in fact any linear map $\tilde{\beta} : T \otimes T \to T \otimes T$ satisfying only the *first two* equalities of (6.20) determines an inverse of (6.22) via (6.23). Now if $\tilde{\beta}$ solves the first two equalities of (6.20) then $\tilde{\tilde{\beta}} := \tilde{\beta} \circ \beta \circ \tilde{\beta}$ solves all of the equalities of (6.20).

While the first two equalities of (6.20) without the third one may have multiple solutions, the solution of the whole set of equalities in (6.20) is unique (provided that it exists). Indeed, if both $\widetilde{\beta}$ and $\widehat{\beta}$ solve the equalities in (6.20) then

$$\widehat{\beta} = \widehat{\beta} \circ \beta \circ \widehat{\beta} = \widehat{\beta} \circ \varXi = \widehat{\beta} \circ \beta \circ \widetilde{\beta} = \underline{\varXi} \circ \widetilde{\beta} = \widetilde{\beta} \circ \beta \circ \widetilde{\beta} = \widetilde{\beta}.$$

Translating the above discussion to the antipode axioms in (6.17) we conclude the uniqueness of the antipode of a weak Hopf algebra. On the other hand, any map $\sigma_0 : T \to T$ which is subject to the first two antipode axioms in Definition 6.20 determines an antipode $a \mapsto \sigma_0(a_{\hat{1}})a_{\hat{2}}\sigma_0(a_{\hat{3}})$.

6.24. The Category of Modules over a Weak (Hopf) Bialgebra. From the characterization of a weak bialgebra T in Theorem 6.19 (ii) as a bialgebroid, we know by Theorems 5.9 and 3.19 that the category $\mathsf{mod}(T)$ of left T-modules admits a monoidal structure lifted from the category of bimodules over the base algebra of the bialgebroid in question; which is obtained now as the image of the map ϵ of (6.10).

Using the separable Frobenius structure of the base algebra $\epsilon(T)$ in Theorem 6.19 (ii), the monoidal product in $\mathsf{mod}(T)$—that is, the $\epsilon(T)$-module tensor product of any T-modules V and W—can be identified with the linear subspace $\{1_{\hat{1}} \cdot v \otimes 1_{\hat{2}} \cdot w \mid v \in V, \ w \in W\}$ of the tensor product vector space $V \otimes W$—see Remark 6.12 and (6.9)—on which T acts by the diagonal action

$$a \otimes 1_{\hat{1}} \cdot v \otimes 1_{\hat{2}} \cdot w \mapsto a_{\hat{1}} \cdot (1_{\hat{1}} \cdot v) \otimes a_{\hat{2}} \cdot (1_{\hat{2}} \cdot w) = a_{\hat{1}} \cdot v \otimes a_{\hat{2}} \cdot w$$

for all $a \in T$, $v \in V$ and $w \in W$. The monoidal unit of $\mathsf{mod}(T)$ is $\epsilon(T)$ with the T-action

$$a \otimes \epsilon(a') \mapsto \epsilon(a\epsilon(a')) = \epsilon(aa')$$

for all $a, a' \in T$; where the last equality follows by part (c) of Exercise 6.16. The associativity constraint is the restriction of the associativity constraint in the category of vector spaces, while the unit constraints in $\mathsf{mod}(T)$ take the respective forms

$$1_{\hat{1}} \cdot \epsilon(a) \otimes 1_{\hat{2}} \cdot v = \epsilon(1_{\hat{1}}a) \otimes 1_{\hat{2}} \cdot v \quad \mapsto \quad \epsilon(1_{\hat{1}}a)1_{\hat{2}} \cdot v = \epsilon(a) \cdot v$$

$$1_{\hat{1}} \cdot v \otimes 1_{\hat{2}} \cdot \epsilon(a) = 1_{\hat{1}} \cdot v \otimes \epsilon(1_{\hat{2}}a) \quad \mapsto \quad \bar{\epsilon}\epsilon(1_{\hat{2}}a)1_{\hat{1}} \cdot v = \bar{\epsilon}\epsilon\bar{\epsilon}(a) \cdot v = \bar{\epsilon}(a) \cdot v$$

for any T-module V, $v \in V$ and $a \in T$ (where the equalities of the second line follow by $\bar{\epsilon} \circ \epsilon = \bar{\epsilon}$ and $\bar{\epsilon} \circ \bar{\epsilon} = \bar{\epsilon}$, see part (b) and a symmetric variant of part (c) of Exercise 6.16).

If T is furthermore a weak Hopf algebra then the corresponding bialgebroid in Theorem 6.19 (ii) is a Hopf algebroid by Theorem 6.22. Hence by Theorems 5.9 and 3.27 the left closed structure of $\mathsf{bim}(\epsilon(T))$ also lifts to $\mathsf{mod}(T)$. Recall from Example 3.24 3 that for any $\epsilon(T)$-bimodule X the internal hom functor $\mathsf{mod}(\epsilon(T)^{\mathsf{op}})(X, -)$ sends an $\epsilon(T)$-bimodule Y to the $\epsilon(T)$-bimodule of the right $\epsilon(T)$-module maps $X \to Y$.

The right $\epsilon(T)$-action on an arbitrary left T-module W is given by $w \otimes \epsilon(a) \mapsto \bar{\epsilon}\epsilon(a) \cdot w = \bar{\epsilon}(a) \cdot w$. Since by a symmetric counterpart of part (h) of Exercise 6.16 the equality $\bar{\epsilon}(a)1_{\hat{1}} \otimes \underline{\epsilon}(1_{\hat{2}}) = 1_{\hat{1}} \otimes \underline{\epsilon}(1_{\hat{2}})\bar{\epsilon}(a)$ holds for all $a \in T$, the vector space $\mathsf{mod}(\epsilon(T)^{\mathsf{op}})(V, W)$ of right $\epsilon(T)$-module maps $V \to W$ can be identified with the linear subspace $\{1_{\hat{1}} \cdot f(\underline{\epsilon}(1_{\hat{2}}) \cdot -) \mid f : V \to W\}$ of the vector space of linear maps $V \to W$, for any T-modules V and W. On this subspace, T acts via the adjoint action

$$a \otimes 1_{\hat{1}} \cdot f(\underline{\epsilon}(1_{\hat{2}}) \cdot -) \mapsto a_{\hat{1}}1_{\hat{1}} \cdot f(\underline{\epsilon}(1_{\hat{2}})\sigma(a_{\hat{2}}) \cdot -) = a_{\hat{1}} \cdot f(\sigma(a_{\hat{2}}) \cdot -)$$

for all $a \in T$ and all linear maps $f : V \to W$. In order to check the last equality, use the second antipode axiom in Definition 6.20; axiom (b) in Definition 6.15; the anti-multiplicativity of σ from Exercise 6.21; the second antipode axiom in Definition 6.20 again; and finally a symmetric variant of Exercise 6.16 (a). They yield

$$1_{\hat{1}} \otimes \underline{\epsilon}(1_{\hat{2}}) = 1_{\hat{1}} \otimes \sigma(1_{\hat{2}})1_{\hat{3}} = 1_{\hat{1}} \otimes \sigma(1_{\hat{1}'}1_{\hat{2}})1_{\hat{2}'}$$
$$= 1_{\hat{1}} \otimes \sigma(1_{\hat{2}})\sigma(1_{\hat{1}'})1_{\hat{2}'} = 1_{\hat{1}} \otimes \sigma(1_{\hat{2}})\underline{\epsilon}(1) = 1_{\hat{1}} \otimes \sigma(1_{\hat{2}}).$$

Together with a further application of the anti-multiplicativity of σ from Exercise 6.21 and of the multiplicativity of the comultiplication, this proves the equality of both stated forms of the adjoint action.

Example 6.25.

1. Let B be a separable Frobenius algebra with counit ψ and comultiplication $b \mapsto b_{\langle 1 \rangle} \otimes b_{\langle 2 \rangle}$ (where Sweedler–Heynemann type implicit summation is understood as in Paragraph 4.2). From Example 5.10 1 and Theorems 6.19 and 6.22 it follows that there is a weak Hopf algebra B^{e}. Its comultiplication $\widehat{\Delta}$, counit $\widehat{\epsilon}$ and antipode σ take the following forms for all $b \otimes b' \in B^{\mathsf{e}}$

$$\widehat{\Delta}(b \otimes b') = (b \otimes 1_{\langle 1 \rangle}) \otimes (1_{\langle 2 \rangle} \otimes b'),$$
$$\widehat{\epsilon}(b \otimes b') = \psi(bb'),$$
$$\sigma(b \otimes b') = b' \otimes 1_{\langle 1 \rangle}\psi(b1_{\langle 2 \rangle}).$$

2. Generalizing Example 4.6 1, consider a category C which has finitely many objects and whose morphisms constitute a set. For any field k, take the vector space $k\mathsf{C}$ spanned by the morphisms of C.

   First we equip $k\mathsf{C}$ with an algebra structure. Define the product of two morphisms f and g to be the composite $f \circ g$ if they are composable (that is, the source of f and the target of g coincide) and zero otherwise. Extending it linearly to $k\mathsf{C}$ in both arguments, we obtain an associative algebra whose unit is $\sum_{X \in \mathsf{C}^0} 1_X$; the sum of the identity morphisms 1_X for all objects X of C (remember that this is a finite sum by assumption).

Observe that the above algebra $k\mathsf{C}$ admits a weak bialgebra structure with comultiplication $\widehat{\Delta}$ and counit $\widehat{\epsilon}$ defined on any morphism f of C as

$$\widehat{\Delta}(f) = f \otimes f \qquad \widehat{\epsilon}(f) = 1$$

and linearly extended to $k\mathsf{C}$. Let us stress that $\widehat{\Delta}$ takes the unit element $\sum_{X \in \mathsf{C}^0} 1_X$ of $k\mathsf{C}$ to $\sum_{X \in \mathsf{C}^0}(1_X \otimes 1_X)$, which differs from $(\sum_{X \in \mathsf{C}^0} 1_X) \otimes (\sum_{X \in \mathsf{C}^0} 1_X)$ unless C has only one object. That is to say, in general $\widehat{\Delta}$ does not preserve the unit element.

As in Theorem 6.19, the above weak bialgebra $k\mathsf{C}$ can be regarded as a bialgebroid over the separable Frobenius algebra $k\mathsf{C}^0$ spanned by the objects of C; with multiplication

$$XY = \begin{cases} X \text{ if } X = Y \\ 0 \text{ if } X \neq Y \end{cases}$$

and comultiplication $X \mapsto X \otimes X$.

If C is in addition a groupoid—that is, all of its morphisms are invertible—then the above weak bialgebra is a weak Hopf algebra with the antipode σ defined on any morphism f of C as $\sigma(f) = f^{-1}$ and linearly extended to $k\mathsf{C}$.

This example extends the well-known fact that the linear spans of monoids are bialgebras and the linear spans of groups are Hopf algebras. In this way it gives an explanation of the origin of the term 'algebroid'.

3. If C is a category with finitely many morphisms, then the algebra of functionals on its set of morphisms (see Example 4.6 2) extends to a weak bialgebra. The comultiplication sends the characteristic functional δ_f of a morphism f in C to the finite sum $\sum_{\{g,h \mid g \circ h = f\}} \delta_g \otimes \delta_h$. The counit sends δ_f to 1 if f is an identity morphism and to 0 otherwise. This weak bialgebra is again a weak Hopf algebra if and only if C is a groupoid; then the antipode sends δ_f to $\delta_{f^{-1}}$.

4. In the algebraic quantum torus of Example 5.10 2 assume the further relation $U^N = 1$—and consistently $q^N = 1$—for some positive integer N which is invertible in k. Then the base algebra B admits a separable Frobenius structure with the comultiplication $U^n \mapsto \frac{1}{N} \sum_{i=1}^{N} U^i \otimes U^{n-i}$ and the counit sending U^n to N if $n = 0$ and to zero otherwise. Hence by Theorem 6.19 there is a corresponding weak bialgebra structure on T_q with the comultiplication and counit

$$\widehat{\Delta}(U^n V^m) = \frac{1}{N} \sum_{i=1}^{N} U^i V^m \otimes U^{n-i} V^m \qquad \widehat{\epsilon}(U^n V^m) = \delta_{n,0} N$$

(in terms of Kronecker's delta symbol δ). By Theorem 6.22 it is a weak Hopf algebra with the antipode sending $U^n V^m$ to $V^{-m} U^n$.

5. In order to describe the symmetry in a concrete physical situation, in the so-called Lee–Yang model, in [18] a 13-dimensional weak Hopf algebra was constructed. As an algebra over the field of complex numbers, it is isomorphic to the direct

sum of the simple algebras of 2×2 and of 3×3 matrices with complex entries. The coalgebra structure is explicitly computed in [18, Section 5].

Remark 6.26. Instead of the left modules over some algebra in Theorem 6.19 (i), one can investigate alternatively the right modules. An argument parallel to the proof of Theorem 6.19 yields a bijection between the following data for any algebra T.

(i) • Monoidal structures on the category of right T-modules and
     • separable Frobenius structures on the forgetful functor to vec.
(ii) • Right bialgebroid structures on T (over some unspecified base algebra) in the sense of [66, Pages 79–80] (see Remark 5.12) and
     • separable Frobenius structures on the base algebra of the bialgebroid.
(iii) Weak bialgebra structures on T.

In particular, combining this with Theorem 6.19 we see that over a separable Frobenius base algebra, both left and right bialgebroids of [66] become equivalent to weak bialgebras.

Remark 6.27. Note that the axioms of a weak (Hopf) bialgebra over a field k are formally self-dual. That is to say, denoting the multiplication and the unit by $\mu : T \otimes T \to T$ and $\eta : k \to T$, denoting the comultiplication and the counit by $\widehat{\Delta} : T \to T \otimes T$ and $\widehat{\epsilon} : T \to k$ and denoting the antipode by $\sigma : T \to T$; and drawing the axioms as commutative diagrams

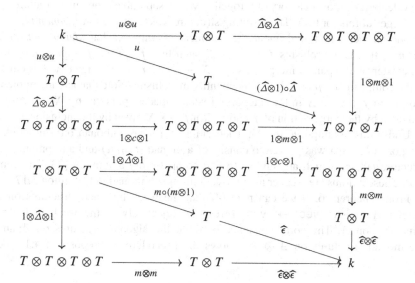

$$T \otimes T \xrightarrow{1 \otimes \sigma} T \otimes T \qquad T \otimes T \xrightarrow{\sigma \otimes 1} T \otimes T \qquad T \otimes T \otimes T \xrightarrow{\sigma \otimes 1 \otimes \sigma} T \otimes T \otimes T$$

$$\Delta \uparrow \qquad \downarrow m \qquad \Delta \uparrow \qquad \downarrow m \qquad (\Delta \otimes 1) \circ \Delta \uparrow \qquad \downarrow m \circ (m \otimes 1)$$

$$T \xrightarrow{\quad \epsilon \quad} T \qquad T \xrightarrow[\epsilon]{\quad \bar{\epsilon} \quad} T \qquad T \xrightarrow{\quad \sigma \quad} T$$

— where $c : T \otimes T \to T \otimes T$ is the flip map $a \otimes a' \mapsto a' \otimes a$—this set of diagrams
is invariant under reversing the arrows and interchanging the roles of the algebra
and the coalgebra structures.

As a consequence of this symmetry, the dual of any valid statement on weak
bialgebras or weak Hopf algebras also holds. For example, dually to Exercise 6.21,
the antipode of a weak Hopf algebra T is a homomorphism from the coalgebra
T to its opposite. As another example, dually to Paragraph 6.24, the category of
comodules over a weak bialgebra is also monoidal, and the category of comodules
over a weak Hopf algebra is left closed.

6.28. Weak Bimonad. Consider a weak bialgebra T. The underlying vector space
induces a functor $T \otimes - : \mathsf{vec} \to \mathsf{vec}$. By Proposition 4.1, the algebra structure of
T determines a monad structure on the functor $T \otimes -$, whose Eilenberg–Moore
category is $\mathsf{mod}(T)$. By Theorem 6.19 it is a monoidal category such that the
forgetful functor to vec admits a separable Frobenius structure.

On the other hand, by Proposition 4.3, the coalgebra structure of T induces an
opmonoidal structure on the functor $T \otimes -$. However, corresponding to the fact that
the algebra and coalgebra structures of T do not constitute a bialgebra—but a weak
bialgebra—the monad and the opmonoidal structures of the functor $T \otimes -$ do not
combine into a bimonad.

Motivated by this, in [23] it was investigated what additional structure on a
monad t on a monoidal category $(\mathsf{A}, I, \otimes)$ is equivalent to a monoidal structure on
the Eilenberg–Moore category of t, together with a separable Frobenius structure on
the forgetful functor to A. The resulting structure was termed a *weak bimonad*.

The answer was found under the technical assumption that *idempotent morphisms*—that is, morphisms $f : X \to X$ such that $f \circ f = f$—*split*; meaning
the existence of a pair of morphisms $(p : X \to Y, i : Y \to X)$ such that f is equal
to the composite $i \circ p$ and $p \circ i$ is the identity morphism. (Note that any idempotent
morphism $f : X \to X$ in the category of vector spaces splits via $p : X \to \mathsf{Im}(f)$
provided by the corestriction of f and $i : \mathsf{Im}(f) \to X$ given by the inclusion.)

Under this assumption, in [23, Theorem 1.5] a weak bimonad on a monoidal
category $(\mathsf{A}, I, \otimes)$ was proven to consist of a monad (t, η, μ) and an opmonoidal
structure (t^0, t^2) which are subject to five compatibility axioms in [23, Theorem
1.5]. These axioms are weaker than those defining a bimonad in Definition 3.17.

From Theorem 6.19 we can read off that the weak bimonad structures on a
functor $A \otimes - : \mathsf{vec} \to \mathsf{vec}$ correspond bijectively to the weak bialgebra
structures on A. The correspondence between the algebra structures on A and
the monad structures on $A \otimes -$ follows the procedure in Proposition 4.1; and

the correspondence between the coalgebra structures on A and the opmonoidal structures on $A \otimes -$ goes as in Proposition 4.3.

Again for a weak bialgebra T, one may consider the same natural transformation β in (4.1); in particular its component (6.18). As we have seen in the proof of Theorem 6.22, it is no longer invertible but satisfies the weaker conditions in (6.20).

In complete analogy with that picture, in [23, Definition 4.1] a *weak Hopf monad* was defined as a weak bimonad—with monad structure (t, η, μ) and opmonoidal structure (t^0, t^2)—whose canonical natural transformation

$$\beta := \quad t(-\otimes t-) \quad \xrightarrow{t^2} \quad t-\otimes tt- \quad \xrightarrow{1\otimes\mu} \quad t-\otimes t-$$

in Definition 3.20 possesses a *weak inverse* $\widetilde{\beta}$. This means that the composites $\beta \circ \widetilde{\beta}$ and $\widetilde{\beta} \circ \beta$ are no longer required to be identity natural transformations but they are required to be equal to certain idempotents canonically associated with the weak bimonad t—the analogues of the idempotent maps of (6.19)—, and satisfy in addition $\widetilde{\beta} \circ \beta \circ \widetilde{\beta} = \widetilde{\beta}$. By [23, Theorem 4.6], a weak bimonad $T \otimes -$ on vec induced by a weak bialgebra T is a weak Hopf monad if and only if T is a weak Hopf algebra.

To begin further reading about weak (Hopf) bialgebras, we recommend e.g. [19, 22, 23, 86, 88, 120].

Chapter 7
(Hopf) Bimonoids in Duoidal Categories

In contrast to the concrete categories—of vector spaces, of bimodules and of bimodules over separable Frobenius algebras, respectively—in the previous chapters, this chapter deals with unspecified *duoidal categories* [3]. These are categories equipped with two compatible monoidal structures. Examples include braided monoidal categories, corresponding to the case when both monoidal structures coincide.

Thanks to the compatibility axioms, monoids with respect to the first monoidal structure, as well as comonoids with respect to the second monoidal structure, constitute monoidal categories. Consequently one can speak about comonoids in this monoidal category of monoids; equivalently, about monoids in this monoidal category of comonoids. They are termed *bimonoids*.

At this level of generality there is no bijection between the morphisms in the category and all natural transformations between the induced functors. So our first task is to identify—with the aid of Lemma 3.7—those monads which are induced by monoids (Proposition 7.6); and those opmonoidal functors which correspond to comonoids (Proposition 7.13). This leads to an equivalence between bimonoids in a duoidal category and bimonads on it of a certain kind (Theorem 7.18).

A characterization of those bimonoids is given whose induced bimonad is a Hopf monad (we avoid calling them Hopf monoids because it is not done so in the literature, but one finds inequivalent candidates for this term; see Paragraph 7.20). Examples of such bimonoids include Hopf monoids in braided monoidal categories—such as the classical Hopf algebras of Chap. 4 and Hopf group algebras of [117]—small groupoids, Hopf algebroids over commutative (but not arbitrary) base algebras in [95], weak Hopf algebras of Chap. 6, Hopf monads of Chap. 3 and certain coalgebra-enriched categories called Hopf categories in [9].

Helpful references should be [3] and [16].

© Springer Nature Switzerland AG 2018

G. Böhm, *Hopf Algebras and Their Generalizations from a Category Theoretical Point of View*, Lecture Notes in Mathematics 2226,
https://doi.org/10.1007/978-3-319-98137-6_7

Definition 7.1 ([74], Section VII.3). A *monoid* in a monoidal category $(\mathsf{A}, I, \otimes)$ is a triple consisting of an object M, a morphism $u : I \to M$—called the *unit*—and a morphism $m : M \otimes M \to M$—called the *multiplication*—for which the following diagrams commute.

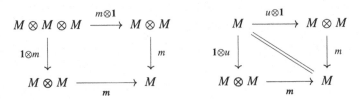

A *morphism of monoids* is a morphism $f : M \to M'$ in A which is compatible with the multiplications and the units in the sense of the following commutative diagrams.

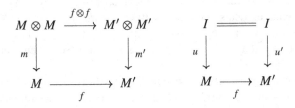

A *comonoid* in a monoidal category is a monoid in the opposite category of Examples 2.2 9 and 3.2 11; that is, a triple $(C, e : C \to I, d : C \to C \otimes C)$ making commutative the same diagrams with reversed arrows. A *morphism of comonoids* is a morphism of monoids in the opposite category (see Example 2.2 9); that is, it renders commutative the same diagrams with reversed vertical arrows.

Example 7.2.

1. The monoidal unit I of any monoidal category $(\mathsf{A}, I, \otimes)$ is a monoid with unit provided by the identity morphism $I \to I$ and multiplication provided by the (equal left and right) unit constraints $I \otimes I \cong I$. Symmetrically, it is a comonoid as well with counit equal to the identity morphism $I \to I$ and comultiplication $I \cong I \otimes I$.

2. A monoid in the monoidal category of sets (see part 1 of Example 3.2) is an ordinary monoid (i.e. a set with an associative multiplication map and unit element).

   On the other hand, any set carries a unique comonoid structure. The counit is a map to the singleton set hence it is unique. Then the counitality condition on any comultiplication $x \mapsto (x', x'')$ enforces $x' = x = x''$.

   More generally, any object X of a Cartesian monoidal category $(\mathsf{A}, I, \times)$ in Example 3.2 1 has a unique comonoid structure as follows. Since the monoidal unit I is terminal, the counit e is the unique morphism $X \to I$.

A comultiplication $d : X \to X \times X$ is counital precisely if the diagonal morphisms in the commutative diagram

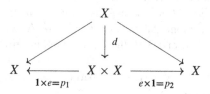

are identities. By the universality of the product $\times$ there is precisely one such morphism d which is evidently coassociative.

Using again that the monoidal unit is terminal, any morphism in **A** is counital. The commutativity of

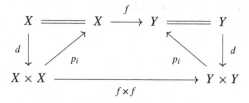

—for any morphism $f : X \to Y$ and for $i = 1, 2$—proves that any morphism f in **A** is comultiplicative; hence a comonoid morphism between the unique comonoid structures on its domain and codomain.

3. A monoid in the monoidal category of vector spaces over a given field k (see part 2 of Example 3.2) is a k-algebra while a comonoid is a k-coalgebra.
4. A monoid in the monoidal category of bimodules over some algebra A (see part 5 of Example 3.2) is an A-ring; that is, an algebra T together with an algebra homomorphism $A \to T$.
5. A monoid in the monoidal category end(**A**) of endofunctors on some category **A** (see part 6 of Example 3.2) is a monad on **A**.
6. A monoid in the monoidal category of coalgebras in Corollary 4.4 is a bialgebra in Definition 4.5. Symmetrically, a comonoid in the monoidal category of algebras in Example 3.2 9 is also a bialgebra.
7. A monoid in the category of $B|B$-corings (for any algebra B) in Corollary 5.6 is a B-bialgebroid in Definition 5.7.

Remark 7.3.

(1)   As in Example 2.5 6, any object M of an arbitrary category **A** can be seen as a functor from the singleton category $\mathbb{1}$ of Example 2.2 1 to **A**. If **A** has a monoidal structure $(I, \otimes)$ and $\mathbb{1}$ is regarded with the monoidal structure in Example 3.2 7, then to give a monoidal structure on the functor $M : \mathbb{1} \to$ **A** is the same thing as to give a monoid structure on the object M.

Consider two monoids (M, u, m) and (M', u', m') in **A**. Any morphism $h : M \to M'$ between their object parts can be seen as a natural transformation from the functor $M : \mathbb{1} \to A$ to $M' : \mathbb{1} \to A$. It is a monoidal natural transformation if and only if h is a monoid morphism. This extends the isomorphism of Exercise 3.14.

(2)   Regard a monoid (M, u, m) in A as a monoidal functor $M : \mathbb{1} \to A$ in part (1); and take a monoidal functor $f : A \to A'$ with nullary part f^0 and binary part f^2. By Exercise 3.10 there is a monoidal structure on the composite

functor $\mathbb{1} \xrightarrow{\ M\ } A \xrightarrow{\ f\ } A'$. Hence by part (1) there is a monoid fM in A' with multiplication and unit

$$fM \otimes' fM \xrightarrow{\ f^2_{M,M}\ } f(M \otimes M) \xrightarrow{\ fm\ } fM \quad \text{and} \quad I' \xrightarrow{\ f^0\ } fI \xrightarrow{\ fu\ } fM.$$

Take a monoid morphism $h : M \to M'$ and regard it as a monoidal natural transformation between the monoidal functors M and $M' : \mathbb{1} \to A$ of part (1). Since the Godement product of monoidal natural transformations is monoidal by Exercise 3.16,

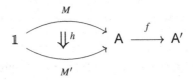

is then a monoidal natural transformation. In view of part (1) this says that a monoidal functor f takes a monoid morphism $h : M \to M'$ to a monoid morphism $fh : fM \to fM'$.

The same reasoning applied to the opposite category of Example 2.2 9 and Example 3.2 11 shows that opmonoidal functors preserve comonoids and their morphisms.

(3)   Consider a monoidal natural transformation φ between monoidal functors f and $f' : A \to B$. Regarding a monoid M in A as a monoidal functor $M : \mathbb{1} \to A$ in part (1), again by Exercise 3.16

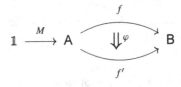

is also a monoidal natural transformation. That is, by part (1), $\varphi_M : fM \to f'M$ is a monoid morphism.

The same reasoning applied again to the opposite category of Examples 2.2 9 and 3.2 11 shows that opmonoidal natural transformations evaluated on comonoids are comonoid morphisms.

(4)   Take an equivalence—with functors $f : \mathsf{A} \to \mathsf{B}$, $g : \mathsf{B} \to \mathsf{A}$ and natural
isomorphisms $\varphi : fg \to 1$, $\psi : gf \to 1$—and a strong monoidal structure on f.
By Exercise 2.19 g can be seen as the right adjoint of f. Hence there is a strong
monoidal structure on g as in Exercise 3.11 with respect to which φ and ψ are
monoidal natural transformations.

Then by part (2) both f and g preserve monoids and monoid morphisms; and
by part (3) ψ_M and φ_N are monoid morphisms for any monoid M in A and
any monoid N in B. This proves that the equivalence (f, g, φ, ψ) induces an
equivalence between the category of monoids in A and the category of monoids
in B.

Example 7.4. By Example 7.2 2 any set X carries a unique comonoid structure in
the Cartesian monoidal category **set** of Example 3.2 1. Since by Remark 7.3 (2) the
strong monoidal 'linear span' functor **set** $\to$ **vec** preserves comonoids, it takes any
set to a coalgebra. The resulting comultiplication sends any basis element $x \in X$ to
$x \otimes x$ and the counit sends x to the number 1.

By Remark 7.3 (2) the 'linear span' functor preserves monoids as well. Namely,
it takes a monoid as in Example 7.2 2 to the monoid algebra in Example 4.6 1.
Together with the coalgebra structure in the previous paragraph they constitute the
'monoid bialgebra' of Example 4.6 1.

Exercise 7.5. Show that a monoid in the monoidal category **alg** of algebras
over a given field k in Example 3.2 9 is precisely a commutative k-algebra.

Now we turn to the study of the relation between monoids in, and monads on, an
arbitrary monoidal category.

Proposition 7.6. *The category of monoids in any monoidal category $(\mathsf{A}, I, \otimes)$ is
equivalent to the following category. The objects consist of a monad (t, η, μ) on
A together with a natural isomorphism $\tau : (t-) \otimes - \to t(- \otimes -)$ such that the
following diagrams commute for all objects X, Y, Z of A.*

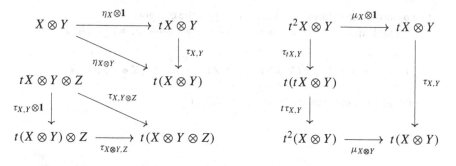

The morphisms $(t, \eta, \mu, \tau) \to (t', \eta', \mu', \tau')$ *are the natural transformations* φ :
$t \to t'$ *for which the following diagrams commute.*

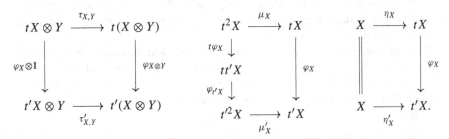

Proof. The category of the claim is the category of monoids in the strict monoidal category of Lemma 3.7. So the stated equivalence is induced by the strong monoidal equivalence of Lemma 3.7 in the way described in Remark 7.3 (4).   □

Remark 7.7. Recall from Lemma 3.7 that in the equivalence of Proposition 7.6 one of the involved natural isomorphisms is provided by the right unit constraint. For particular objects (t, η, μ, τ) in which the functor t is of the form $T \otimes -$ for some object T of A, and τ is the associativity constraint with fixed first argument T, the other involved natural isomorphism in the equivalence of Proposition 7.6 is also induced by the right unit constraint (as derived from the triangle axiom).

Hence if in a monoidal category $(A, I, \otimes)$ the right unit constraint is trivial, then the equivalence of Proposition 7.6 yields a bijection between the monoid structures on any object T in A and those monad structures (η, μ) on the functor $T \otimes - : A \to A$ for which $\eta_{X \otimes Y} = \eta_X \otimes 1$ and $\mu_{X \otimes Y} = \mu_X \otimes 1$ for all objects X, Y of A. In this sense Proposition 7.6 extends Proposition 4.1.

The following definition is quoted from [3, Definition I.6.1] where it was termed a *2-monoidal category*. The more commonly used term *duoidal category* was proposed in [107]. In a more restrictive form—when the morphisms ξ_0 and ξ_0^0 below are invertible—it appeared in [7].

Definition 7.8. A *duoidal structure* on a category A consists of

- *two* monoidal structures $(I, \diamond)$ and $(J, \blacklozenge)$ on the category A
- a monoidal structure on the functor $\diamond : (A, J, \blacklozenge) \times (A, J, \blacklozenge) \to (A, J, \blacklozenge)$

$$(J \xrightarrow{\xi^0} J \diamond J , \quad (X \diamond Y) \blacklozenge (V \diamond Z) \xrightarrow{\xi_{X,Y,V,Z}} (X \blacklozenge V) \diamond (Y \blacklozenge Z))$$

- a monoidal structure on the functor $I : (\mathbb{1}, !, !) \to (A, J, \blacklozenge)$

$$(J \xrightarrow{\xi_0^0} I , \quad I \blacklozenge I \xrightarrow{\xi_0} I)$$

such that the (not explicitly denoted) associativity and unit constraints of the monoidal category $(A, I, \diamond)$ are monoidal natural transformations.

Equivalently,

- *two* monoidal structures $(I, \diamond)$ and $(J, \bullet)$ on the category A
- an opmonoidal structure on the functor $\bullet : (\mathsf{A}, I, \diamond) \times (\mathsf{A}, I, \diamond) \to (\mathsf{A}, I, \diamond)$

$$(\ I \bullet I \xrightarrow{\xi_0} I \ , \quad (X \diamond Y) \bullet (V \diamond Z) \xrightarrow{\xi_{X,Y,V,Z}} (X \bullet V) \diamond (Y \bullet Z) \ )$$

- an opmonoidal structure on the functor $J : (\mathbb{1}, !, !) \to (\mathsf{A}, I, \diamond)$

$$(\ J \xrightarrow{\xi_0^0} I \ , \quad J \xrightarrow{\xi^0} J \diamond J \ )$$

such that the (not explicitly denoted) associativity and unit constraints of the monoidal category $(\mathsf{A}, J, \bullet)$ are opmonoidal natural transformations.

Exercise 7.9. Spell out the diagrams which the structure morphisms $\xi, \xi^0, \xi_0, \xi_0^0$ of a duoidal category $(\mathsf{A}, I, \diamond, J, \bullet)$ must render commutative.

Exercise 7.10. Show that in Definition 7.8 the morphism $\xi_0^0 : J \to I$ is redundant; (omitting the unit constraints) it can be written either as

$$J = (I \diamond J) \bullet (J \diamond I) \xrightarrow{\xi} (I \bullet J) \diamond (J \bullet I) = I \qquad \text{or}$$

$$J = (J \diamond I) \bullet (I \diamond J) \xrightarrow{\xi} (J \bullet I) \diamond (I \bullet J) = I.$$

Example 7.11.

1. A *braiding* [62, Definition 2.1] [74, Section XII.1] on a monoidal category $(\mathsf{A}, K, \otimes)$ is an invertible natural transformation c between $\otimes : \mathsf{A} \times \mathsf{A} \to \mathsf{A}$ and its opposite $\mathsf{A} \times \mathsf{A} \xrightarrow{\text{flip}} \mathsf{A} \times \mathsf{A} \xrightarrow{\otimes} \mathsf{A}$—with components $c_{X,Y} : X \otimes Y \to Y \otimes X$ for arbitrary objects X, Y of A—such that (omitting the associativity constraints) the following diagrams commute for all objects X, Y and Z.

$$
\begin{array}{ccc}
X \otimes Y \otimes Z & \xrightarrow{c_{X,Y \otimes Z}} & Y \otimes Z \otimes X \\
{\scriptstyle c_{X,Y} \otimes 1} \searrow & & \nearrow {\scriptstyle 1 \otimes c_{X,Z}} \\
& Y \otimes X \otimes Z &
\end{array}
\qquad
\begin{array}{ccc}
& X \otimes Z \otimes Y & \\
{\scriptstyle 1 \otimes c_{Y,Z}} \nearrow & & \searrow {\scriptstyle c_{X,Z} \otimes 1} \\
X \otimes Y \otimes Z & \xrightarrow{c_{X \otimes Y,Z}} & Z \otimes X \otimes Y
\end{array}
$$

A braiding c is called a *symmetry* if $c_{Y,X} \circ c_{X,Y}$ is the identity morphism $X \otimes Y \to X \otimes Y$ for all objects X and Y.

Some examples of braided monoidal categories are the following.

(a) The monoidal category of sets in part 1 of Example 3.2 is a symmetric monoidal category with symmetry provided by the flip maps $x \otimes y \mapsto y \otimes x$.
(b) The monoidal category of vector spaces in part 2 of Example 3.2 is also a symmetric monoidal category with symmetry provided by the flip maps.
(c) The monoidal category of modules over a commutative ring A in Example 3.2 4 is a symmetric monoidal category too, with symmetry provided by the flip maps.
(d) The monoidal category hil of Hilbert spaces in Example 3.2 8 is also symmetric via the extension of the symmetry on vector spaces (that is, the flip map).
(e) Consider the category $\mathsf{vec}^{\mathbb{Z}}$ of vector spaces graded by the additive group of integers. It has a monoidal structure as in Example 3.2 3. Any invertible element u of the base field defines a braiding on it, with components $\{X_n\}_{n \in \mathbb{Z}} \otimes \{Y_n\}_{n \in \mathbb{Z}} \to \{Y_n\}_{n \in \mathbb{Z}} \otimes \{X_n\}_{n \in \mathbb{Z}}$ induced by the maps $X_i \otimes Y_j \to Y_j \otimes X_i$, $x \otimes y \mapsto u^{i+j} y \otimes x$. This braiding is a symmetry if and only if $u^2 = 1$.
(f) The monoidal category $\mathsf{disc}(X)$ of Example 3.2 7, for some monoid X, is symmetric if and only if X is Abelian, otherwise it is not even braided.

However, the monoidal category of bimodules in part 5 of Example 3.2, and the monoidal category of endofunctors in part 6 of Example 3.2 are not braided in general.

Any braided monoidal category $(\mathsf{A}, K, \otimes, c)$ can be regarded as a duoidal category [3, Section I.6.3] with equal monoidal structures $(I, \diamond) = (K, \otimes) = (J, \bullet)$ and

$$\xi_0^0 : K \xrightarrow{\ 1\ } K \qquad \xi_0 : K \otimes K \xrightarrow{\ \cong\ } K$$

$$\xi^0 : K \xrightarrow{\ \cong\ } K \otimes K \qquad \xi : X \otimes Y \otimes V \otimes Z \xrightarrow{\ 1 \otimes c_{Y,V} \otimes 1\ } X \otimes V \otimes Y \otimes Z.$$

In fact, a certain converse also holds. If in a duoidal category $(\mathsf{A}, I, \diamond, J, \bullet)$ all structure morphisms ξ, ξ_0, ξ^0 and ξ_0^0 are invertible then both monoidal categories $(\mathsf{A}, I, \diamond)$ and $(\mathsf{A}, J, \bullet)$ are braided and they are isomorphic as braided monoidal categories, see [3, Proposition 6.11].

2. By [3, Example 6.19] any monoidal category $(\mathsf{A}, J, \bullet)$ which has binary products and a terminal object can be regarded as a duoidal category with $\diamond$-monoidal structure the Cartesian one in Example 3.2 1.

This applies in particular to the category $\mathsf{span}(X)$ of *spans* over an arbitrary set X [3, Example I.6.17]. Its objects are triples consisting of a set A and two maps $s, t : A \to X$. Such an object can be visualized as a directed graph with vertex set X and edge set A; the maps s and t taking the edges to their source and target, respectively. The morphisms in $\mathsf{span}(X)$ are maps $f : A \to A'$ which are compatible with the maps to X in the sense of the following commutative diagram.

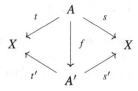

This category $\mathsf{span}(X)$ is monoidal via the *pullback* of spans

$$X \xleftarrow{\quad (p,q) \mapsto t'(p) \quad} A' \blacklozenge A := \{(p,q) \in A' \times A \,|\, s'(p) = t(q)\} \xrightarrow{\quad (p,q) \mapsto s(q) \quad} X$$

—that is, the set of pairs of "consecutive edges" in A and A'—with monoidal unit $X \xleftarrow{\;1\;} X \xrightarrow{\;1\;} X$.

Furthermore, it also has the *binary product* of spans

$$X \xleftarrow{\quad (p,q) \mapsto t'(p) = t(q) \quad} A' \diamond A := \{(p,q) \in A' \times A \,|\, t'(p) = t(q),\ s'(p) = s(q)\} \xrightarrow{\quad (p,q) \mapsto s'(p) = s(q) \quad} X$$

—that is, the set of pairs of "parallel edges" in A and A'—with monoidal unit the terminal object $X \xleftarrow{\quad (x,y) \mapsto x \quad} X \times X \xrightarrow{\quad (x,y) \mapsto y \quad} X$.

These monoidal structures combine into a duoidal category whose structure morphisms are the following.

$\xi_0^0 : X \to X \times X,$ $\qquad\qquad x \mapsto (x, x)$

$\xi_0 : (X \times X) \blacklozenge (X \times X) \to X \times X,$ $\qquad\qquad (x, y = x', y') \mapsto (x, y')$

$\xi^0 : X \to X \cong X \diamond X,$ $\qquad\qquad x \mapsto x$

$\xi : (A \diamond B) \blacklozenge (A' \diamond B') \to (A \blacklozenge A') \diamond (B \blacklozenge B'),$
$$((a, b), (a', b')) \mapsto ((a, a'), (b, b'))$$

pictorially:

$$\begin{pmatrix} \overset{a}{\leftarrow} \\ \overset{b}{\leftarrow} \end{pmatrix} \begin{pmatrix} \overset{a'}{\leftarrow} \\ \overset{b'}{\leftarrow} \end{pmatrix} \xmapsto{\;\xi\;} \begin{pmatrix} (\overset{a}{\leftarrow} \overset{a'}{\leftarrow}) \\ (\overset{b}{\leftarrow} \overset{b'}{\leftarrow}) \end{pmatrix}$$

3. For a *commutative* algebra B, the category $\mathsf{bim}(B)$ of B-bimodules admits the following duoidal structure [3, Example I.6.18]. The first monoidal product is the usual B-module tensor product

$$V \diamond W := V \otimes W / \{v \cdot b \otimes w - v \otimes b \cdot w \mid b \in B, \ v \in V, \ w \in W\}$$

whose unit is B with left and right actions provided by the multiplication; see Example 3.2 5. Using that B is commutative, we can regard any B-bimodule as a $B \otimes B$-module and the second monoidal product can be defined as the $B \otimes B$-module tensor product

$$V \blacklozenge W := V \otimes W / \{b \cdot v \cdot b' \otimes w - v \otimes b \cdot w \cdot b' \mid b, b' \in B, \ v \in V, \ w \in W\}$$

whose unit is $B \otimes B$ with left action provided by the multiplication in the first factor and right action provided by the multiplication in the second factor. The structure morphisms take the following forms.

$\xi_0^0 : B \otimes B \to B,$ $\qquad\qquad\qquad\qquad b \otimes b' \mapsto bb'$

$\xi_0 : B \blacklozenge B \to B,$ $\qquad\qquad\qquad\qquad b \blacklozenge b' \mapsto bb'$

$\xi^0 : B \otimes B \to (B \otimes B) \diamond (B \otimes B),$ $\qquad b \otimes b' \mapsto (b \otimes 1) \diamond (1 \otimes b')$

$\xi : (X \diamond Y) \blacklozenge (V \diamond W) \to (X \blacklozenge V) \diamond (Y \blacklozenge W),$
$$(x \diamond y) \blacklozenge (v \diamond w) \mapsto (x \blacklozenge v) \diamond (y \blacklozenge w)$$

4. For a *separable Frobenius* algebra B—with counit ψ and comultiplication $b \mapsto b_{\langle 1 \rangle} \otimes b_{\langle 2 \rangle}$, where the Sweedler–Heynemann implicit summation index notation of Paragraph 4.2 is used—the category of B^e bimodules admits the following duoidal structure [25, Section 4]. The monoidal product $V \blacklozenge W$ of B^e-bimodules V and W is their usual B^e-module tensor product

$$V \otimes_{B^e} W := V \otimes W / \{v \cdot (b \otimes b') \otimes w - v \otimes (b \otimes b') \cdot w \mid b, b' \in B, \ v \in V, \ w \in W\},$$

whose monoidal unit is B^e with actions provided by the multiplication; see Example 3.2 5. The other monoidal product $V \diamond W$ is a twisted B^e-module tensor product

$$V \otimes W / \{(b \otimes 1) \cdot v \cdot (b' \otimes 1) \otimes w - v \otimes (1 \otimes \psi(1_{\langle 1 \rangle} b) 1_{\langle 2 \rangle}) \cdot w \cdot (1 \otimes b') \mid$$
$$b, b' \in B, \ v \in V, \ w \in W\}$$

(where the so-called *Nakayama automorphism* $b \mapsto \psi(1_{\langle 1 \rangle} b) 1_{\langle 2 \rangle}$ of the separable Frobenius algebra B occurs). Its monoidal unit is B^e with suitably twisted actions. The structure morphisms $\xi_0^0, \xi^0, \xi_0, \xi$ are given in terms of the separable Frobenius algebra structure of B, see [25, Theorem 4.3].

7.12. Duoidal Categories as Intercategories. *Intercategories* [56] are three-dimensional category-like structures. That is, they have one kind of object, three kinds of arrows, three kinds of faces and one kind of three-dimensional cell. There are three compatible compositions, one of which is strictly associative and unital; the others are associative and unital up to coherent isomorphisms.

As discussed in [57, Section 2], duoidal categories can be seen as intercategories in which there is only one object, there are only identity arrows of all kinds and only identity faces of two kinds. The third kind of faces play the role of the objects in the duoidal category and the three-dimensional cells play the role of its morphisms. The strictly associative and unital composition of the intercategory serves as the composition of the morphisms of the duoidal category and the other two compositions yield both monoidal products.

The next subject of our investigation is the relation between comonoids in, and opmonoidal endofunctors on, a monoidal category $(A, I, \diamond)$ underlying a duoidal category $(A, I, \diamond, J, \bullet)$. We will see a nice interplay between both monoidal structures.

Proposition 7.13. *For any duoidal category $(A, I, \diamond, J, \bullet)$, the following categories are equivalent.*

(i) *The category of comonoids in the monoidal category $(A, I, \diamond)$.*

(ii) *The category whose objects consist of an opmonoidal functor (t, t^0, t^2) : $(A, I, \diamond) \to (A, I, \diamond)$ and an opmonoidal natural isomorphism*

$$
\begin{array}{ccc}
(A, I, \diamond) \times (A, I, \diamond) & \xrightarrow{\quad (t,t^0,t^2) \times 1 \quad} & (A, I, \diamond) \times (A, I, \diamond) \\[4pt]
{\scriptstyle (\bullet, \xi_0, \xi)} \Big\downarrow & \quad \Downarrow \tau \quad & \Big\downarrow {\scriptstyle (\bullet, \xi_0, \xi)} \\[4pt]
(A, I, \diamond) & \xrightarrow{\quad (t,t^0,t^2) \quad} & (A, I, \diamond)
\end{array}
$$

such that

$$
\begin{array}{ccc}
tX \bullet Y \bullet Z & \xrightarrow{\ \tau_{X,Y} \bullet 1\ } & t(X \bullet Y) \bullet Z \\[4pt]
 & \searrow{\scriptstyle \tau_{X,Y \bullet Z}} & \Big\downarrow {\scriptstyle \tau_{X \bullet Y, Z}} \\[4pt]
 & & t(X \bullet Y \bullet Z)
\end{array}
\tag{7.1}
$$

commutes for all objects X, Y, Z of A; and whose morphisms are the opmonoidal natural transformations $\varphi : t \to t'$ rendering commutative the following diagram.

$$
\begin{array}{ccc}
tX \bullet Y & \xrightarrow{\ \tau_{X,Y}\ } & t(X \bullet Y) \\[2pt]
{\scriptstyle \varphi_X \bullet 1}\big\downarrow & & \big\downarrow{\scriptstyle \varphi_X \bullet Y} \\[2pt]
t'X \bullet Y & \xrightarrow[\ \tau'_{X,Y}\ ]{} & t'(X \bullet Y)
\end{array}
$$

Proof. This is somewhat analogous to Lemma 3.7.

Regard a comonoid (C, e, d) in $(\mathsf{A}, I, \diamond)$ as an opmonoidal functor $C :$ $(\mathbb{1}, !, !) \to (\mathsf{A}, I, \diamond)$ as in Remark 7.3 (1). By the desired functor f from the category of part (i) to the category of part (ii) it is sent to the composite opmonoidal functor

$$
(\mathsf{A}, I, \diamond) \xrightarrow{\ (C,e,d) \times 1\ } (\mathsf{A}, I, \diamond) \times (\mathsf{A}, I, \diamond) \xrightarrow{\ (\bullet,\xi_0,\xi)\ } (\mathsf{A}, I, \diamond)
$$

—the nullary and binary parts of its opmonoidal structure are computed as in Exercise 3.10 and come out as

$$
C \bullet I \xrightarrow{\ e \bullet 1\ } I \bullet I \xrightarrow{\ \xi_0\ } I \quad \text{and} \quad C \bullet (X \diamond Y) \xrightarrow{\ d \bullet 1\ } (C \diamond C) \bullet (X \diamond Y) \xrightarrow{\ \xi\ } (C \bullet X) \diamond (C \bullet Y),
$$
$$
\tag{7.2}
$$

respectively—supplemented with the appropriate components of the opmonoidal associativity natural isomorphism of $(\mathsf{A}, J, \bullet)$.

A comonoid morphism $h : (C, e, d) \to (C', e', d')$—seen as an opmonoidal natural transformation between the opmonoidal functors C and $C' : (\mathbb{1}, !, !) \to$ $(\mathsf{A}, I, \diamond)$ as in Remark 7.3 (1)—is sent by f to the opmonoidal natural transformation

$$
(\mathsf{A}, I, \diamond) \underset{\substack{\big\Downarrow {\scriptstyle h \times 1}}}{\overset{(C,e,d)\times 1}{\underset{(C',e',d')\times 1}{\rightrightarrows}}} (\mathsf{A}, I, \diamond) \times (\mathsf{A}, I, \diamond) \xrightarrow{\ (\bullet,\xi_0,\xi)\ } (\mathsf{A}, I, \diamond).
$$

The opmonoidal functor $(J, \xi_0^0, \xi^0) : (\mathbb{1}, !, !) \to (\mathsf{A}, I, \diamond)$ can be regarded as a comonoid in $(\mathsf{A}, I, \diamond)$, see Remark 7.3 (1). Our functor g from the category of part (ii) to the category of part (i) sends an object (t, t^0, t^2, τ) to the image of the comonoid (J, ξ_0^0, ξ^0) under the opmonoidal functor (t, t^0, t^2); that is, the comonoid

$$(tJ, \quad tJ \xrightarrow{\ t\xi_0^0\ } tI \xrightarrow{\ t^0\ } I \ , \quad tJ \xrightarrow{\ t\xi^0\ } t(J \diamond J) \xrightarrow{\ t_{J,J}^2\ } tJ \diamond tJ \ ).$$

A morphism $\varphi : (t, t^0, t^2, \tau) \to (t', t'^0, t'^2, \tau')$ is sent by g to the comonoid morphism $\varphi_J : tJ \to t'J$ (it is indeed a comonoid morphism by Remark 7.3 (3).)

Applying gf to a comonoid (C, e, d) in part (i) we re-obtain (C, e, d) up to the omitted right unit constraint of $(\mathsf{A}, J, \bullet)$.

Applying fg to an object (t, t^0, t^2, τ) of the category in part (ii) we get the opmonoidal functor $tJ \bullet -$ with the nullary part

$$tJ \bullet I \xrightarrow{\ t\xi_0^0 \bullet 1\ } tI \bullet I \xrightarrow{\ t^0 \bullet 1\ } I \bullet I \xrightarrow{\ \xi_0\ } I$$

and binary part

$$tJ \bullet (X \diamond Y) \xrightarrow{\ t\xi^0 \bullet 1\ } t(J \diamond J) \bullet (X \diamond Y) \xrightarrow{\ t_{J,J}^2 \bullet 1\ } (tJ \diamond tJ) \bullet (X \diamond Y) \xrightarrow{\ \xi\ } (tJ \bullet X) \diamond (tJ \bullet Y)$$

and the natural isomorphism provided by the (omitted) associativity constraint of $(\mathsf{A}, J, \bullet)$. A natural isomorphism from this to (t, t^0, t^2, τ) is given by the components $\tau_{J,X} : tJ \bullet X \to tX$ for any object X (omitting again the left unit constraint). It is opmonoidal by the opmonoidality of τ and it is a morphism in the category of part (ii) by (7.1). $\square$

Remark 7.14. For an object (t, t^0, t^2, τ) of the category of part (ii) of Proposition 7.13 in which the functor t is of the form $C \bullet -$ for some object C; and τ is provided by the associativity constraint with fixed first argument C, the natural isomorphism $fg(t) = tJ \bullet - \to t$ in the last paragraph of the proof of Proposition 7.13 is obtained by applying the right unit constraint in the first factor of $(C \bullet J) \bullet - \to C \bullet -$.

The other natural isomorphism $gf \to 1$ in the proof of Proposition 7.13 is the right unit constraint for every object. Hence whenever the right unit constraint is trivial, Proposition 7.13 yields a bijection between the comonoid structures on the object C and those opmonoidal structures (t^0, t^2) on the functor $C \bullet -$ for which the associativity constraint with fixed first argument C becomes opmonoidal; that is, the following diagrams commute.

$$
\begin{array}{ccc}
C \bullet I \bullet I & \xrightarrow{\ t^0 \bullet 1\ } & I \bullet I \\
{\scriptstyle 1 \bullet \xi^0} \downarrow & & \downarrow {\scriptstyle \xi^0} \\
C \bullet I & \xrightarrow[\ t^0\ ]{} & I
\end{array}
\qquad
\begin{array}{ccc}
C \bullet (X \diamond Y) \bullet (V \diamond W) & \xrightarrow{\ t_{X,Y}^2 \bullet 1\ } & ((C \bullet X) \diamond (C \bullet Y)) \bullet (V \diamond W) \\
{\scriptstyle 1 \bullet \xi} \downarrow & & \downarrow {\scriptstyle \xi} \\
C \bullet ((X \bullet V) \diamond (Y \bullet W)) & \xrightarrow[\ t_{X \bullet V, Y \bullet W}^2\ ]{} & (C \bullet X \bullet V) \diamond (C \bullet Y \bullet W)
\end{array}
$$

$$(7.3)$$

Regard the symmetric monoidal category vec of vector spaces over a given field k as a duoidal category in the way described in Example 7.11 1. Then the first diagram of (7.3) reduces to a trivial identity and the second one holds for any natural transformation $t^2 : C \otimes - \otimes - \to C \otimes - \otimes C \otimes -$ by its form in Exercise 2.8: it must be induced by a linear map $C \to C \otimes C$, $c \mapsto c_1 \otimes c_2$ (where implicit summation is understood as in Paragraph 4.2); that is, each component $t^2_{X,Y} : C \otimes X \otimes Y \to C \otimes X \otimes C \otimes Y$ must send any rank one element $c \otimes x \otimes y$ to $c_1 \otimes x \otimes c_2 \otimes y$.

This shows that Proposition 7.13 extends Proposition 4.3.

Corollary 7.15. *For any duoidal category* $(A, I, \diamond, J, \bullet)$, *the category of comonoids in* $(A, I, \diamond)$ *admits the following monoidal structure. The monoidal unit is J with the comultiplication ξ^0 and counit ξ^0_0. The monoidal product of any comonoids (C, e, d) and (C', e', d') is $C \bullet C'$ with the comultiplication and counit*

$$C \bullet C' \xrightarrow{\;d \bullet d'\;} (C \diamond C) \bullet (C' \diamond C') \xrightarrow{\;\xi\;} (C \bullet C') \diamond (C \bullet C') \qquad C \bullet C' \xrightarrow{\;e \bullet e'\;} I \bullet I \xrightarrow{\;\xi_0\;} I.$$

The monoidal product of any comonoid morphisms f and g is $f \bullet g$.

Proof. The opmonoidal functor (J, ξ^0_0, ξ^0) from the singleton category to $(A, I, \diamond)$ can be seen as a comonoid in $(A, I, \diamond)$ as in Remark 7.3 (1).

For a comonoid (C, e, d), the functor $C \bullet - : (A, I, \diamond) \to (A, I, \diamond)$ is opmonoidal by Proposition 7.13 with nullary and binary parts in (7.2). Hence by Remark 7.3 (2) it takes the comonoid (C', e', d') to the stated comonoid $C \bullet C'$.

If evaluated at any comonoids, the opmonoidal associativity and unit constraints of the monoidal category $(A, J, \bullet)$ are comonoid morphisms, see Remark 7.3 (3). Finally, $f \bullet g$ is a comonoid morphism for any comonoid morphisms f and g being the image of the comonoid morphism $f \times g$ under the opmonoidal functor $\bullet$, see Remark 7.3 (2).   □

Remark 7.16. If a category A admits a duoidal structure $(I, \diamond, J, \bullet)$ then it is immediate from the symmetry of the axioms that the opposite category A^{op} in Example 2.2 9 inherits the duoidal structure $(J, \bullet, I, \diamond)$ (with the same structure morphisms ξ^0_0 and ξ; and the roles of ξ^0 and ξ_0 interchanged).

Applying Corollary 7.15 to the duoidal category $(A^{op}, J, \bullet, I, \diamond)$ we conclude that the comonoids in the monoidal category $(A^{op}, J, \bullet)$—that is, the monoids in $(A, J, \bullet)$—also constitute a monoidal category with the monoidal product $\diamond$.

Definition 7.17. [3, Section I.6.5.1] A *bimonoid* in a duoidal category $(A, I, \diamond, J, \bullet)$ is a monoid in the monoidal category of comonoids in $(A, I, \diamond)$ described in Corollary 7.15. Equivalently, it is a comonoid in the monoidal category of monoids in $(A, J, \bullet)$ explained in Remark 7.16. Explicitly, this means an object T equipped with the structures of a monoid (T, u, m) in $(A, J, \bullet)$ and of a comonoid (T, e, d) in $(A, I, \diamond)$ such that the following diagrams commute.

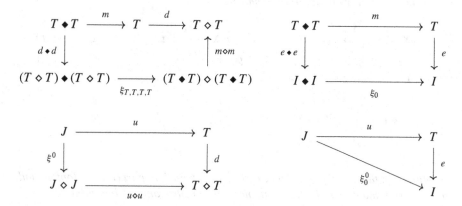

A *morphism of bimonoids* is a monoid morphism in the category of comonoids in $(A, I, \diamond)$; equivalently, a comonoid morphism in the category of monoids in $(A, J, \bullet)$; equivalently, a morphism in A which is a monoid morphism in $(A, J, \bullet)$ as well as a comonoid morphism in $(A, I, \diamond)$.

Theorem 7.18. *For any duoidal category $(A, I, \diamond, J, \bullet)$ the following categories are equivalent.*

(i) The category of bimonoids in $(A, I, \diamond, J, \bullet)$.

(ii) The category whose objects consist of an opmonoidal monad t on the monoidal category $(A, I, \diamond)$—with monad structure (η, μ) and opmonoidal structure (t^0, t^2)—together with an opmonoidal natural isomorphism $\tau : (t-) \bullet - \rightarrow t(- \bullet t)$ such that the following diagrams commute.

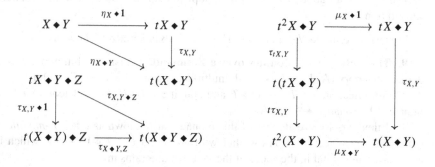

The morphisms are the opmonoidal natural transformations $\varphi : t \to t'$ *for which the following diagrams commute.*

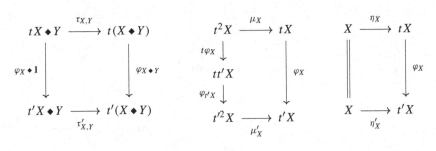

Moreover, the stated equivalence takes a bimonoid T in part (i) to a Hopf monad in part (ii) if and only if the following natural transformation is invertible

$$T \bullet (X \diamond (T \bullet Y)) \xrightarrow{d \bullet 1} (T \diamond T) \bullet (X \diamond (T \bullet Y)) \xrightarrow{\xi_{T,T,X,T \bullet Y}} (T \bullet X) \diamond (T \bullet T \bullet Y) \xrightarrow{1 \diamond (m \bullet 1)} (T \bullet X) \diamond (T \bullet Y) \cdot$$
$$(7.4)$$

Proof. Variants of this result can be found in [27, Theorem 6.7] and [3, Proposition 6.39].

The category of part (ii) of Proposition 7.13 is strict monoidal with the same monoidal product in Lemma 3.7, see Exercises 3.10 and 3.16. The category of part (ii) of the current claim is the category of monoids therein.

The equivalence functor f in Proposition 7.13 is strong monoidal with nullary part of the opmonoidal structure provided by the right unit constraint, and binary part given by the associativity constraint of $(A, J, \bullet)$ (much like as in Lemma 3.7). Hence it induces the stated equivalence by Remark 7.3 (4).

The final claim about $T \bullet -$ being a Hopf monad follows by a straightforward substitution. $\square$

In view of Remarks 7.7 and 7.14, Theorem 7.18 generalizes Theorem 4.8.

7.19. The Category of Modules over a Bimonoid. Consider a bimonoid T in a duoidal category $(A, I, \diamond, J, \bullet)$ with multiplication $m : T \bullet T \to T$, unit $u : J \to T$, comultiplication $d : T \to T \diamond T$ and counit $e : T \to I$. By Theorem 7.18 it induces a bimonad $T \bullet -$ on $(A, I, \diamond)$.

An Eilenberg–Moore algebra of this monad—also known as a left T-module—consists of an object V of A together with an *action* $v : T \bullet V \to V$ which is associative and unital in the sense of the first two diagrams in

$$
\begin{array}{ccc}
T \bullet T \bullet V & \xrightarrow{m \bullet 1} & T \bullet V \\
{\scriptstyle 1 \bullet v}\downarrow & & \downarrow{\scriptstyle v} \\
T \bullet V & \xrightarrow{v} & V
\end{array}
\qquad
\begin{array}{ccc}
V & \xrightarrow{u \bullet 1} & T \bullet V \\
& \searrow & \downarrow{\scriptstyle v} \\
& & V
\end{array}
\qquad
\begin{array}{ccc}
T \bullet V & \xrightarrow{1 \bullet f} & T \bullet V' \\
{\scriptstyle v}\downarrow & & \downarrow{\scriptstyle v'} \\
V & \xrightarrow{f} & V'.
\end{array}
$$

A morphism of Eilenberg–Moore algebras is a morphism $f : V \to V'$ in $\mathbf{A}$ for which the third diagram above commutes.

By Theorem 3.19 this category of T-modules has a monoidal structure lifted from $(\mathbf{A}, I, \diamond)$. Thus the monoidal unit is I with the action

$$T \bullet I \xrightarrow{\ e \bullet 1\ } I \bullet I \xrightarrow{\ \xi_0\ } I$$

induced by the nullary part of the opmonoidal structure. The monoidal product of any two T-modules $v : T \bullet V \to V$ and $w : T \bullet W \to W$ is $V \diamond W$ with the *diagonal action*

$$T \bullet (V \diamond W) \xrightarrow{\ d \bullet 1\ } (T \diamond T) \bullet (V \diamond W) \xrightarrow{\ \xi\ } (T \bullet V) \diamond (T \bullet W) \xrightarrow{\ v \diamond w\ } V \diamond W$$

induced by the binary part of the opmonoidal structure.

By Theorem 3.27, whenever the monoidal category $(\mathbf{A}, I, \diamond)$ is left closed, the lifting of its internal hom to the above monoidal category of T-modules is equivalent to the invertibility of (7.4).

7.20. Competing Notions of Hopf Monoid. It looks tempting to term a bimonoid in a duoidal category a *Hopf monoid* whenever the corresponding natural transformation (7.4) is invertible. However, this is not the only definition in the literature.

Consider a bimonoid T in a duoidal category $(\mathbf{A}, I, \diamond, J, \bullet)$, with multiplication $m : T \bullet T \to T$, unit $u : J \to T$, comultiplication $d : T \to T \diamond T$ and counit $e : T \to I$.

1. In [107, Section 4.7, Definition] and [27, Definition 6.7] invertibility of the component of the natural transformation (7.4) at the objects $X = J = Y$ is assumed; that is, invertibility of the single morphism

$$T \bullet (J \diamond T) \xrightarrow{\ d \bullet 1\ } (T \diamond T) \bullet (J \diamond T) \xrightarrow{\ \xi\ } T \diamond (T \bullet T) \xrightarrow{\ 1 \diamond m\ } T \diamond T.$$

2. By a *Hopf module* over T we mean a triple consisting of an object X of $\mathbf{A}$ together with an action $\underline{x} : T \bullet X \to X$ of the monad $T \bullet -$ and a coaction $\overline{x} : X \to T \diamond X$ of the comonad $T \diamond -$ such that the following diagram commutes.

$$
\begin{array}{ccccc}
T \bullet X & \xrightarrow{\ \underline{x}\ } & X & \xrightarrow{\ \overline{x}\ } & T \diamond X \\
{\scriptstyle d \bullet \overline{x}} \downarrow & & & & \uparrow {\scriptstyle m \diamond \underline{x}} \\
(T \diamond T) \bullet (T \diamond X) & & \xrightarrow{\hspace{3cm}\xi\hspace{3cm}} & & (T \bullet T) \diamond (T \bullet X)
\end{array}
$$

A morphism of Hopf modules is a morphism in A which commutes with the actions of the monad $T \blacklozenge -$ and also with the coactions of the comonad $T \diamond -$.

Since (J, ξ_{50}^0, ξ^0) is a comonoid in $(\mathsf{A}, I, \diamond)$—see Corollary 7.15—its image under the strong monoidal functor in Example 3.6 5 is a comonad $J \diamond -$ on A. There is a canonical functor k from its Eilenberg–Moore category to the category of Hopf modules over any bimonoid T. The functor k sends an object $y : Y \to J \diamond Y$ to the Hopf module with the object part $T \blacklozenge Y$, with action of $T \blacklozenge -$ and coaction of $T \diamond -$ in

$$T \blacklozenge T \blacklozenge Y \xrightarrow{m \blacklozenge 1} T \blacklozenge Y \qquad T \blacklozenge Y \xrightarrow{d \blacklozenge y} (T \diamond T) \blacklozenge (J \diamond Y) \xrightarrow{\xi} T \diamond (T \blacklozenge Y),$$

respectively. It sends a morphism $f : Y \to Y'$ to $1 \blacklozenge f : T \blacklozenge Y \to T \blacklozenge Y'$. The question of when this functor k is an equivalence was investigated in [24, Theorem 3.11] and [2].

3. The unit of any monoid is a monoid morphism. Following [24], the unit of the particular monoid underlying the bimonoid T is said to be a *Galois extension* $J \to T$ by T if for any left T-module $v : T \blacklozenge V \to V$ the morphism

$$T \blacklozenge (J \diamond V) \xrightarrow{d \blacklozenge 1} (T \diamond T) \blacklozenge (J \diamond V) \xrightarrow{\xi} T \diamond (T \blacklozenge V) \xrightarrow{1 \diamond v} T \diamond V$$

is invertible.

4. The bimonoid T not only induces the bimonad $T \blacklozenge -$ studied so far, but also a second opmonoidal monad $- \blacklozenge T$ on $(\mathsf{A}, I, \diamond)$ with corresponding natural transformation (3.4) taking the form

$$((X \blacklozenge T) \diamond Y) \blacklozenge T \xrightarrow{1 \blacklozenge d} ((X \blacklozenge T) \diamond Y) \blacklozenge (T \diamond T) \xrightarrow{\xi} (X \blacklozenge T \blacklozenge T) \diamond (Y \blacklozenge T) \xrightarrow{(1 \blacklozenge m) \diamond 1} (X \blacklozenge T) \diamond (Y \blacklozenge T).$$

Its invertibility was investigated in [24].

There are even more possibilities. The bimonoid T also induces monoidal comonads $T \diamond -$ and $- \diamond T$ on $(\mathsf{A}, J, \blacklozenge)$ with respective canonical natural transformations

$$(T \diamond X) \blacklozenge (T \diamond Y) \xrightarrow{1 \blacklozenge (d \diamond 1)} (T \diamond X) \blacklozenge (T \diamond T \diamond Y) \xrightarrow{\xi} (T \blacklozenge T) \diamond (X \blacklozenge (T \diamond Y)) \xrightarrow{m \diamond 1} T \diamond (X \blacklozenge (T \diamond Y))$$

$$(X \diamond T) \blacklozenge (Y \diamond T) \xrightarrow{(1 \diamond d) \blacklozenge 1} (X \diamond T \diamond T) \blacklozenge (Y \diamond T) \xrightarrow{\xi} ((X \diamond T) \blacklozenge Y) \diamond (T \blacklozenge T) \xrightarrow{1 \diamond m} ((X \diamond T) \blacklozenge Y) \diamond T.$$

The invertibility of this last one was analyzed in [25].

For a classical bialgebra T over a field all of these properties are equivalent to the invertibility of (7.4) and hence to T being a Hopf algebra. There is no reason to expect, however, their equivalence for any bimonoid T in an arbitrary duoidal category. Their precise relation has been analyzed using the language of monoidal bicategories.

Situations when the condition of part 1 implies the invertibility of (7.4) for all objects X and Y were described in [27, Theorem 7.10].

In [16] the properties of parts 2 and 3 were proven to be equivalent to the invertibility of the natural transformation (7.4) for bimonoids in a large class of duoidal categories—including braided monoidal categories in Example 7.11 1, the category of spans over a given set in Example 7.11 2, the category of bimodules over a commutative algebra in Example 7.11 3, and the category of bimodules over the enveloping algebra of a separable Frobenius algebra in Example 7.11 4.

In principle the conditions of part 4 are inequivalent to the others and to Definition 3.20. Still, as we shall see in Example 7.21, in concrete examples they may turn out to be equivalent.

Example 7.21.

1. Regard a braided monoidal category $(\mathsf{A}, K, \otimes, c)$ as a duoidal category, as described in part 1 of Example 7.11. A bimonoid in this duoidal category reduces to the usual notion [98, Section I.6.1.3] of a bimonoid in the braided monoidal category $(\mathsf{A}, K, \otimes, c)$. It consists of

   • an object T of A
   • a monoid structure $(u : K \to T, m : T \otimes T \to T)$ in $(\mathsf{A}, K, \otimes)$
   • a comonoid structure $(e : T \to K, d : T \to T \otimes T)$ in $(\mathsf{A}, K, \otimes)$

   such that the following diagrams commute.

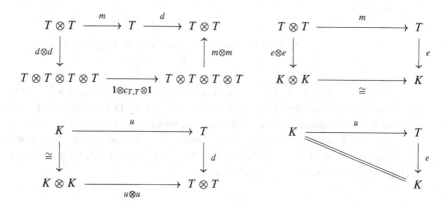

By Theorem 7.18, any bimonoid T in the above sense induces a bimonad $T \otimes -$ on $(\mathsf{A}, K, \otimes)$. Consequently, the monoidal structure $(K, \otimes)$ of A lifts to the Eilenberg–Moore category of the monad $T \otimes -$ on A. It results in the monoidal unit $e : T \to K$; and the monoidal product of Eilenberg–Moore algebras $v : T \otimes V \to V$ and $w : T \otimes W \to W$ given by the object $V \otimes W$ with the *diagonal action*

$$T \otimes V \otimes W \xrightarrow{d \otimes 1 \otimes 1} T \otimes T \otimes V \otimes W \xrightarrow{1 \otimes c \otimes 1} T \otimes V \otimes T \otimes W \xrightarrow{v \otimes w} V \otimes W.$$

The bimonad $T \otimes -$ on $(\mathsf{A}, K, \otimes)$ induced by a bimonoid T is a Hopf monad if and only if

$$\beta_{X,Y} := T \otimes X \otimes T \otimes Y \xrightarrow{d \otimes 1 \otimes 1 \otimes 1} T \otimes T \otimes X \otimes T \otimes Y \xrightarrow{1 \otimes c_{T,X} \otimes 1 \otimes 1} T \otimes X \otimes T \otimes T \otimes Y$$

$$\xrightarrow{1 \otimes 1 \otimes m \otimes 1} T \otimes X \otimes T \otimes Y$$

is an isomorphism, for any objects X and Y. The natural transformation β renders commutative the diagram

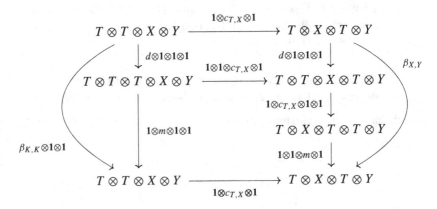

(Note that the component $c_{T,K}$ of the braiding (not) occurring in the left column is trivial. Indeed, putting in the first axiom of braiding in Example 7.11 1 the objects Y and Z equal to the monoidal unit K, we obtain $c_{X,K} \circ c_{X,K} = c_{X,K}$ for all objects X. Since $c_{X,K}$ is invertible by definition, this proves $c_{X,K} = 1$ modulo the omitted associativity and unit constraints.) From this we infer that $T \otimes -$ is a Hopf monad on $(\mathsf{A}, K, \otimes)$ if and only if

$$\beta_{K,K} = T \otimes T \xrightarrow{d \otimes 1} T \otimes T \otimes T \xrightarrow{1 \otimes m} T \otimes T$$

is invertible. In complete analogy with Theorem 4.8 (see [16, Section 8.1]), this is equivalent to the existence of an *antipode* morphism $z : T \to T$ for which the following diagram commutes.

$$\begin{array}{ccccc}
T & \xrightarrow{d} & T \otimes T & \xrightarrow{1 \otimes z} & T \otimes T \\
\downarrow{\scriptstyle d} & \searrow{\scriptstyle e} & K & & \downarrow{\scriptstyle m} \\
& & \downarrow{\scriptstyle u} & & \\
T \otimes T & \xrightarrow{z \otimes 1} & T \otimes T & \xrightarrow{m} & A
\end{array}$$

Indeed, if $\beta_{K,K}$ is invertible then the antipode z is constructed as the composite morphism

$$T \xrightarrow{1 \otimes u} T \otimes T \xrightarrow{\beta_{K,K}^{-1}} T \otimes T \xrightarrow{e \otimes 1} T.$$

Conversely, if there is an antipode morphism z then $\beta_{K,K}$ has the inverse

$$T \otimes T \xrightarrow{d \otimes 1} T \otimes T \otimes T \xrightarrow{1 \otimes z \otimes 1} T \otimes T \otimes T \xrightarrow{1 \otimes m} T \otimes T$$

and these constructions are mutual inverses. A bimonoid equipped with a (necessarily unique) antipode is called a *Hopf monoid*.

This class of examples includes the classical *bialgebras* and *Hopf algebras* in Chap. 4; which are bimonoids and Hopf monoids, respectively, in the symmetric monoidal category of vector spaces (see part (b) of Example 7.11 1). More generally, it includes *bialgebras* and *Hopf algebras* over any commutative ring A which are bimonoids and Hopf monoids, respectively, in the symmetric monoidal category of A-modules (see part (c) of Example 7.11 1). It also includes semigroups and groups which are bimonoids and Hopf monoids, respectively, in the symmetric monoidal category of sets (see part (a) Example 7.11 1). It includes $\mathbb{Z}$-*graded bialgebras* and $\mathbb{Z}$-*graded Hopf algebras* which are bimonoids and Hopf monoids, respectively, in the braided monoidal category of vector spaces graded by the additive group of integers $\mathbb{Z}$ (see part (e) of Example 7.11 1). Let us mention a final, somewhat more exotic example introduced in [117] and motivated by physics, in connection with homotopy quantum field theory. For any monoid G, monoidal functors T from the monoidal category disc(G) of Example 3.2 7 to the monoidal category of coalgebras in Corollary 4.4 were termed in [32, Definition 1.8] *semi-Hopf G-algebras*. They can be seen as bimonoids in a symmetric monoidal category, the so-called *Zunino category* in [32, Section 2.2]. Such a bimonoid is a Hopf monoid if and only if T is a *Hopf G-algebra* in the sense of [117]; that is, it possesses an antipode in the sense of [32, Definition 1.8].

2. Let us take the duoidal category $\mathsf{span}(X)$ of part 2 of Example 7.11 for an arbitrary set X. Following [3, Example I.6.43], its bimonoids can be described as follows.

Since $(\mathsf{span}(X), X \times X, \diamond)$ is a Cartesian monoidal category, every object has a unique comonoid structure and any morphism is a morphism of comonoids, see Example 7.2 2. (More explicitly, from an arbitrary object (A, t, s) there is a unique map of spans to $X \times X$, the one sending a to $(t(a), s(a))$. Hence there is a unique counital comultiplication $A \to A \diamond A$: the diagonal one sending a to (a, a).) This amounts to saying that a bimonoid in $\mathsf{span}(X)$ is the same thing as a monoid in $(\mathsf{span}(X), X, \blacklozenge)$.

A monoid in $(\mathsf{span}(X), X, \blacklozenge)$, on the other hand, is precisely a small category with the object set X. Indeed, for any span $s, t : A \to X$, an element a of A can be seen as a morphism $s(a) \to t(a)$. The unit of the monoid A is a map $\mathbf{1} : X \to A$. It is a map of spans; that is, it sends any $x \in X$ to a morphism $1_x : x \to x$. The multiplication is a map $\circ : A \blacklozenge A \to A$. It is a map of spans; that is, it sends a pair of morphisms $f : x \to y$ and $g : y \to z$ to a morphism $g \circ f : x \to z$. The associativity and unitality axioms of monoids reduce precisely to the diagrams of Definition 2.1.

By Theorem 7.18 any bimonoid A in $\mathsf{span}(X)$—that is, any small category A with the object set X—induces a bimonad $\mathsf{A} \blacklozenge -$ on $(\mathsf{span}(X), X \times X, \diamond)$. It is a Hopf monad if and only if A is a groupoid; that is, every morphism in A is invertible. Indeed, the component of the natural transformation (7.4) at arbitrary spans P and Q over X is a map from

$$\mathsf{A} \blacklozenge (P \diamond (\mathsf{A} \blacklozenge Q)) = \{(a, p, a', q) \in \mathsf{A} \times P \times \mathsf{A} \times Q|$$
$$s(a) = t(p) = t(a'), \ s(a') = t(q), \ s(p) = s(q)\}$$

to

$$(\mathsf{A} \blacklozenge P) \diamond (\mathsf{A} \blacklozenge Q) = \{(a, p, a', q) \in \mathsf{A} \times P \times \mathsf{A} \times Q|$$
$$s(a) = t(p), \ s(a') = t(q), \ t(a) = t(a'), \ s(p) = s(q)\},$$

it sends (a, p, a', q) to $(a, p, a \circ a', q)$. If every morphism a has an inverse then this map possesses the inverse sending (a, p, a', q) to $(a, p, a^{-1} \circ a', q)$. Conversely, if it is invertible for all spans P and Q then it is invertible in particular for $P = X \times X$ and $Q = X$; that is,

$$\mathsf{A} \blacklozenge \mathsf{A} \to \{(a, a') \in \mathsf{A} \times \mathsf{A}|t(a) = t(a')\}, \qquad (a, a') \mapsto (a, a \circ a')$$

is invertible. If this is the case then its inverse must send $(a, 1_{t(a)})$ for any morphism a in A to the pair whose first member is a and the second member provides the inverse of A—see [24, Corollary 4.6] or [25, Proposition 8.2].

The map sending a morphism of A to its inverse plays the role of a generalized antipode map. Note that it is not a span map from $X \xleftarrow{t} A \xrightarrow{s} X$ to itself; but to the 'opposite span' $X \xleftarrow{s} A \xrightarrow{t} X$.

In [24] and [25], respectively, the invertibility of all morphisms in the category A was proven to be equivalent to some other conditions in part 4 of Paragraph 7.20.

3. For a commutative algebra B, consider the duoidal category $\mathrm{bim}(B)$ in part 3 of Example 7.11. As discussed in [3, Example I.6.44], a monoid in $(\mathrm{bim}(B), B \otimes B, \bullet)$ is a $B \otimes B$-ring A with unit homomorphism $B \otimes B \to A$ landing in the center; while a comonoid in $(\mathrm{bim}(B), B, \diamond)$ is a so-called B-coring. A bimonoid in the duoidal category $\mathrm{bim}(B)$ is precisely a B-bialgebroid T such that the unit of the $B \otimes B$-ring T; that is, the algebra homomorphism $\eta : B \otimes B \to T$ lands in the center of T.

By Theorem 5.9 an arbitrary B-bialgebroid T induces a bimonad on $\mathrm{bim}(B)$ with underlying functor $T \boxtimes -$ as in Paragraph 5.1. For the particular class of B-bialgebroids whose map η lands in the center, it is the same bimonad $T \bullet -$ induced by T as in Theorem 7.18. Consequently, by Theorem 5.9, $T \bullet -$ is a Hopf monad if and only if T is a Hopf algebroid. In [24, Proposition 4.9] this was proven to hold if and only if T is a Hopf algebroid in the sense of [95]. This means the existence of a linear map $\sigma : T \to T$—the *antipode*—such that

$$\sigma(c\eta(b \otimes 1)) = \sigma(c)\eta(1 \otimes b) \quad \sigma(c\eta(1 \otimes b)) = \sigma(c)\eta(b \otimes 1)$$
$$c_1\sigma(c_2) = \eta(\epsilon(c) \otimes 1) \qquad \sigma(c_1)c_2 = \eta(1 \otimes \epsilon(c))$$

for all $b \in B$ and $c \in T$, where $c_1 \otimes_B c_2$ denotes the image of c under the comultiplication, with implicit summation understood as in Paragraph 4.2; and ϵ stands for the counit of the B-bialgebroid T.

Note that the antipode is not a B-bimodule map from T to itself but to the 'opposite bimodule' T with the actions $B \otimes T \otimes B \to T, b \otimes t \otimes b' \mapsto b' \cdot t \cdot b$.

4. Take next a separable Frobenius algebra B and the duoidal category of B^{e}-bimodules in part 4 of Example 7.11. It was proven in [25, Theorem 4.11] that the bimonoids therein are precisely the weak bialgebras whose base algebra is B.

By Theorem 6.19 any weak bialgebra T over the base algebra B can be seen as a B-bialgebroid. Hence by Theorem 5.9 it induces a bimonad on $(\mathrm{bim}(B), B, \otimes_B)$ by taking the B^{e}-module tensor product $T \boxtimes -$ in Paragraph 5.1. Note the difference between this bimonad and the bimonad $T \bullet -$ on a different category $(\mathrm{bim}(B^{\mathrm{e}}), I, \diamond)$ induced by T as in Theorem 7.18.

For the bimonad $T \bullet -$ the component of (7.4) at any B^{e}-bimodules X and Y sends $a \bullet (x \diamond (a' \bullet y))$ to $(a_{\hat{1}} \bullet x) \diamond (a_{\hat{2}}a' \bullet y)$ (where multiplication is denoted by juxtaposition and for the comultiplication the same Sweedler–Heynemann implicit summation index notation is used as in the proof of Theorem 6.19). Its invertibility can be investigated by the same method as in [25, Proposition 8.3]. If T is a weak Hopf algebra with the antipode σ then it has the (well-defined) inverse sending $(a \blacklozenge x) \diamond (a' \blacklozenge y)$ to $a_{\hat{1}} \blacklozenge (x \diamond (\sigma(a_{\hat{2}})a' \blacklozenge y))$. Conversely, if (7.4) is invertible for any B^e-bimodules X and Y, then it is invertible, in particular, by choosing both X and Y to be the B^e-bimodule $B^{\otimes 4}$ with the action

$$(b \otimes c) \otimes (p \otimes q \otimes p' \otimes q') \otimes (b' \otimes c') \mapsto bp \otimes qc \otimes p'b' \otimes c'q'.$$

In terms of the map ω of (6.22), this component differs from $\omega \otimes 1 \otimes 1 \otimes 1$ by the isomorphisms

$$T \blacklozenge (B^{\otimes 4} \diamond (T \blacklozenge B^{\otimes 4})) \cong T1_{\hat{1}} \otimes \bar{\varepsilon}(1_{\hat{2}})T \otimes B \otimes B \otimes B$$

and

$$(T \blacklozenge B^{\otimes 4}) \diamond (T \blacklozenge B^{\otimes 4}) \cong 1_{\hat{1}}T \otimes 1_{\hat{2}}T \otimes B \otimes B \otimes B.$$

Hence it is invertible if and only if the map of (6.22) is invertible; equivalently, by Theorem 6.19, T is a weak Hopf algebra. This proves that the bimonad $T \blacklozenge -$ on $(\mathsf{bim}(B^e), I, \diamond)$ induced by a weak bialgebra T over the base algebra B is a Hopf monad if and only if T is a weak Hopf algebra. Note that the antipode is not a B^e-bimodule map from T to itself but to some twisted bimodule.

In [25, Proposition 8.3] the same property—of T being a weak Hopf algebra—was shown to be equivalent to another condition in part 4 of Paragraph 7.20.

5. For any category A there is a monoidal category $\mathsf{cat}(\mathsf{A}^{\mathsf{op}} \times \mathsf{A}, \mathsf{set})$ whose objects are the functors from $\mathsf{A}^{\mathsf{op}} \times \mathsf{A}$ to the category of sets—the so-called *profunctors* $\mathsf{A} \nrightarrow \mathsf{A}$—and whose morphisms are the natural transformations. The monoidal structure is recalled e.g. in [16, Section 5.5].

If A is a monoidal category then $\mathsf{cat}(\mathsf{A}^{\mathsf{op}} \times \mathsf{A}, \mathsf{set})$ has a duoidal structure; see [107, Section 4.3, Example 5] or [16, Paragraphs 5.5 and 3.3]. The opmonoidal monads t on the monoidal category A correspond bijectively to the bimonoids of the form $\mathsf{A}(-, t(-))$ in the duoidal category $\mathsf{cat}(\mathsf{A}^{\mathsf{op}} \times \mathsf{A}, \mathsf{set})$; see again [107, Section 4.6 Observations 1] or [16, Paragraphs 5.5 and 3.3]. At this level of generality, invertibility of the corresponding natural transformation (7.4) is not known to correspond to the existence of any kind of antipode.

An object L in a monoidal category $(\mathsf{A}, I, \otimes)$ is the *left dual* of an object R; equivalently, R is the *right dual* of L, if there are some morphisms $L \otimes R \to I$ and $I \to R \otimes L$—called the *evaluation* and the *coevaluation*, respectively—satisfying the evident variants of the triangle conditions in (2.2). Under the additional assumption that in a monoidal category A every object has a left dual and a right dual, invertibility of the natural transformation (7.4) for a bimonad t on A—seen as a bimonoid $\mathsf{A}(-, t(-))$ in the duoidal category $\mathsf{cat}(\mathsf{A}^{\mathsf{op}} \times \mathsf{A}, \mathsf{set})$—is equivalent to the existence of an antipode for t in the sense of [29]; see [16, Paragraph 8.5]. This antipode is a natural transformation from the profunctor $\mathsf{A}(-, t(-)) : \mathsf{A} \nrightarrow \mathsf{A}$ to the profunctor t^- in [16, Paragraph 8.5] (not to itself).

6. A category *enriched* in a monoidal category $(\mathsf{V}, I, \otimes)$ consists of

- a class of objects $X, Y, \ldots$
- for each pair of objects X, Y an *object* $\mathsf{A}(X, Y)$ of V
- for each object X a *morphism* $\mathbf{1} : I \to \mathsf{A}(X, X)$ in V
- for each triple of objects X, Y, Z a *morphism* $\circ : \mathsf{A}(Y, Z) \otimes \mathsf{A}(X, Y) \to \mathsf{A}(X, Z)$ in V

such that for all objects X, Y, Z, V the diagrams of Definition 2.1 commute—replacing in them $\times$ by $\otimes$. (Thus a small category is in fact a category enriched in the Cartesian monoidal category of sets.)

In [14, Paragraph 4.9] a duoidal category was constructed in which a category T, enriched in the monoidal category of coalgebras in Corollary 4.4, can be seen as a bimonoid. For this bimonoid the natural transformation (7.4) is invertible if and only if T is a *Hopf category*; that is, it possesses an antipode in the sense of [9, Definition 3.3]. Once again, the antipode of a Hopf category connects two different objects of the duoidal category.

Remark 7.22. In most of the cases discussed in Example 7.21 we saw that the bimonad induced by a certain bimonoid is a Hopf monad if and only if some kind of antipode exists. We should stress that this is *not a general feature* of bimonoids in a duoidal category; there are induced bimonads which are Hopf monads although they are not known to possess any kind of antipode; see e.g. part 5 of Example 7.21. In general there is not even a natural candidate for the target object of a potential antipode morphism.

In fact, the existence of certain antipode morphisms in some of the instances of Example 7.21 follows from the fact that these examples belong to bimonoids in a distinguished class of duoidal categories, discussed in [16] in the framework of monoidal bicategories.

For further reading about duoidal categories and in particular about (Hopf) bimonoids in them, we recommend the relevant chapters of the monograph [3] and the papers [8, 14, 16, 24, 26, 27, 35, 107].

Chapter 8
Solutions to the Exercises

Exercise 2.8. Consider functors $W \otimes_A -$ and $W' \otimes_A - : \mathrm{mod}(A) \to \mathrm{mod}(B)$ induced by B-A bimodules W and W' as in part 4 of Example 2.5. Show that any natural transformation between them is induced by a unique B-A bimodule map $W \to W'$ as in part 3 of Example 2.7.

Solution. To a natural transformation $\varphi : W \otimes_A - \to W' \otimes_A -$ associate the left B-module map

$$W \xrightarrow{\ \cong\ } W \otimes_A A \xrightarrow{\ \varphi_A\ } W' \otimes_A A \xrightarrow{\ \cong\ } W'. \tag{8.1}$$

Right multiplication by any fixed element a of A defines a left A-module map $A \to A$. Naturality of φ with respect to it yields the commutative diagram

$$
\begin{array}{ccccccc}
W & \xrightarrow{\ \cong\ } & W \otimes_A A & \xrightarrow{\ \varphi_A\ } & W' \otimes_A A & \xrightarrow{\ \cong\ } & W' \\
{\scriptstyle -\cdot a}\downarrow & & {\scriptstyle 1\otimes_A -\cdot a}\downarrow & & \downarrow{\scriptstyle 1\otimes_A -\cdot a} & & \downarrow{\scriptstyle -\cdot a} \\
W & \xrightarrow[\ \cong\ ]{} & W \otimes_A A & \xrightarrow[\ \varphi_A\ ]{} & W' \otimes_A A & \xrightarrow[\ \cong\ ]{} & W'
\end{array}
$$

that is, right A-linearity of the map of (8.1).

Starting with a B-A bimodule map and iterating the construction in part 3 of Example 2.7 and the above one, we obviously re-obtain the original bimodule map. Starting with a natural transformation $\varphi : W \otimes_A - \to W' \otimes_A -$ and iterating these constructions in the opposite order, we re-obtain φ by its naturality with respect to

© Springer Nature Switzerland AG 2018

G. Böhm, *Hopf Algebras and Their Generalizations from a Category Theoretical Point of View*, Lecture Notes in Mathematics 2226, https://doi.org/10.1007/978-3-319-98137-6_8

the left A-module map $A \to V$, $a \mapsto a \cdot v$, induced by an arbitrary fixed element v of any left A-module V. That is, by commutativity of the following diagram.

$$
\begin{array}{ccccc}
W & \xrightarrow{\;\cong\;} & W \otimes_A A & \xrightarrow{\;1 \otimes_A - \cdot v\;} & W \otimes_A V \\
(8.1) \downarrow & & \downarrow {\scriptstyle \varphi_A} & & \downarrow {\scriptstyle \varphi_V} \\
W' & \xrightarrow[\;\cong\;]{} & W' \otimes_A A & \xrightarrow[\;1 \otimes_A - \cdot v\;]{} & W' \otimes_A V
\end{array}
$$

Exercise 2.10. Prove that the operations of Paragraph 2.9 obey the following interchange law. For any functors $f, f', f'' : \mathsf{A} \to \mathsf{B}$ and $g, g', g'' : \mathsf{B} \to \mathsf{C}$, and for any natural transformations $\varphi : f \to f'$, $\varphi' : f' \to f''$, $\gamma : g \to g'$, $\gamma' : g' \to g''$, the equality $(\gamma' \circ \gamma)(\varphi' \circ \varphi) = (\gamma'\varphi') \circ (\gamma\varphi)$ holds.

Solution. The following diagram commutes by the naturality of γ.

$$
\begin{array}{ccccc}
gfX & \xrightarrow{\;g\varphi_X\;} & gf'X & \xrightarrow{\;g\varphi'_X\;} & gf''X \\
 & {\scriptstyle \gamma_{f'X}} \downarrow & & \downarrow {\scriptstyle \gamma_{f''X}} & \\
 & g'f'X & \xrightarrow[\;g'\varphi'_X\;]{} & g'f''X & \xrightarrow[\;\gamma'_{f''X}\;]{} & g''f''X
\end{array}
$$

Its top path is the component of $(\gamma' \circ \gamma)(\varphi' \circ \varphi)$ at an arbitrary object X and the bottom path is the component of $(\gamma'\varphi') \circ (\gamma\varphi)$ at X.

Exercise 2.17. Show that whenever a functor possesses a (left or right) adjoint, it is unique up to a natural isomorphism; and this isomorphism is compatible with the unit and the counit of the adjunction.

Solution. Assume that both l and l' are left adjoints of the same functor r; with respective units $\eta : 1 \to rl$ and $\eta' : 1 \to rl'$ and respective counits $\varepsilon : lr \to 1$ and $\varepsilon' : l'r \to 1$. Consider the natural transformation

$$
\vartheta := l \xrightarrow{\;1\eta'\;} lrl' \xrightarrow{\;\varepsilon 1\;} l'.
$$

By the triangle conditions (2.2) on both adjunctions and the naturality of ε and η, respectively, the diagrams

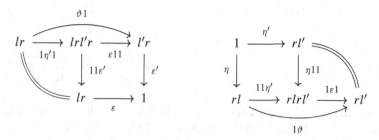

commute (this is what is meant by the compatibility of ϑ with the units and counits). Then the diagrams

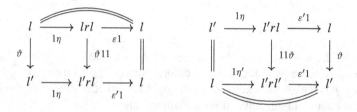

also commute by the triangle conditions (2.2) and by the naturality of ϑ and ε', respectively. This proves that ϑ has the inverse $\varepsilon'1 \circ 1\eta$.

Exercise 2.18. For any adjunction $l \dashv r : A \to B$ with unit η and counit ε, verify bijective correspondences between natural transformations of the following kinds.

(1) Between natural transformations $fr \to g$ and $f \to gl$, for any functors $f : B \to C, g : A \to C$ and any category C.
(2) Between natural transformations $f \to rg$ and $lf \to g$, for any functors $f : C \to B, g : C \to A$ and any category C.

Solution. Take a natural transformation $\varphi : fr \to g$ as in part (1) and associate to it $f \xrightarrow{1\eta} frl \xrightarrow{\varphi 1} gl$. Conversely, to $\psi : f \to gl$ associate the composite $fr \xrightarrow{\psi 1} glr \xrightarrow{1\varepsilon} g$. These constructions are mutual inverses by the commutativity

of both diagrams

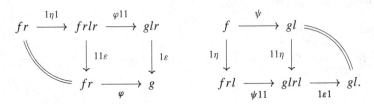

The triangular regions on the left of the first diagram and on the right of the second diagram commute by the triangle conditions of (2.2). Both squares commute by naturality of φ and ψ, respectively.

There is a symmetric bijection in part (2) provided by the mutually inverse maps

$$f \xrightarrow{\varphi} rg \mapsto lf \xrightarrow{1\varphi} lrg \xrightarrow{\varepsilon 1} g \qquad\qquad lf \xrightarrow{\psi} g \mapsto f \xrightarrow{\eta 1} rlf \xrightarrow{1\psi} rg.$$

Exercise 2.19. Prove that if the functors $f : A \to B$ and $g : B \to A$ take part in an equivalence (with the natural isomorphisms $\varphi : fg \to 1$ and $\psi : gf \to 1$) then f is both the left adjoint and the right adjoint of g; moreover, the unit and the counit of both adjunctions are isomorphisms.

Solution. The family of maps

$$A(X, gY) \xrightarrow{f} B(fX, fgY) \xrightarrow{B(fX,\varphi Y)} B(fX, Y) , \qquad h \mapsto \varphi Y \circ fh \qquad (8.2)$$

is natural in the objects X of A and Y of B by the functoriality of f and naturality of φ. We claim that each member has the inverse

$$B(fX, Y) \xrightarrow{B(fX,\varphi Y^{-1})} B(fX, fgY) \xrightarrow{g} A(gfX, gfgY) \xrightarrow{A(\psi X^{-1},\psi gY)} A(X, gY) \qquad (8.3)$$

sending a morphism k to $\psi gY \circ g\varphi Y^{-1} \circ gk \circ \psi X^{-1}$.

Taking $h \in A(X, gY)$ and applying to it the above maps (8.2) and (8.3), we re-obtain h by the naturality of ψ. This proves in particular that f is faithful; and a symmetric argument shows that so is g.

In order to see that the composite of (8.3) and (8.2) in the opposite order gives the identity map too, note that the following diagram commutes by the naturality of ψ for any morphism $p : fX \to fX'$.

$$gfX \xrightarrow{gf\psi_X^{-1}} gfgfX \xrightarrow{gfgp} gfgfX' \xrightarrow{gf\psi_{X'}} gfX'$$

$$\psi_X \downarrow \qquad \psi_{gfX} \downarrow \qquad \qquad \downarrow \psi_{gf'X} \qquad \downarrow \psi_{X'}$$

$$X \xrightarrow{\psi_X^{-1}} gfX \xrightarrow{gp} gfX' \xrightarrow{\psi_{X'}} X'$$

Since $\psi_{X'}$ is invertible and g is faithful, this proves $p = f\psi_{X'} \circ fgp \circ f\psi_X^{-1}$. Applying it to $p = \varphi_Y^{-1} \circ k$ we see that the top part of

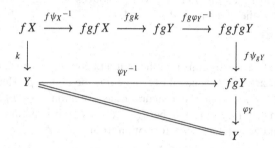

commutes for all $k \in B(fX, Y)$, proving that the application of (8.3) and next (8.2) to k yields k as the result.

From (8.2) we read off the form $\varphi_Y : fgY \to Y$ of the counit; and from (8.3) we read off the form $\psi_{gfX} \circ g\varphi_{fX}^{-1} \circ \psi_X^{-1} : X \to gfX$ of the unit. Both of them are composites of isomorphisms hence isomorphisms themselves.

This proves that f is the left adjoint of g with invertible unit and counit. Interchanging the roles of f and g we see that f is also the right adjoint of g with invertible unit and counit.

A (slightly different) proof of this claim can be found in [74, Section IV.4 Theorem 1].

Exercise 2.25. For any monad (t, η, μ) on some category A consider the so-called *Kleisli category* A_t; whose objects are the same objects of A and whose morphisms $X \nrightarrow Y$ are the morphisms $X \to tY$ in A. Such morphisms are composed by the rule

$$(Y \xrightarrow{g} Z) \circ (X \xrightarrow{f} Y) := (X \xrightarrow{f} tY \xrightarrow{tg} t^2Z \xrightarrow{\mu_Z} tZ)$$

and the identity morphism $X \nrightarrow X$ is $\eta_X : X \to tX$.

(continued)

(1) Show that the functor

$$f_t : A \to A_t, \qquad (X \xrightarrow{h} Y) \mapsto (X \xrightarrow{h} Y \xrightarrow{\eta Y} tY)$$

is the left adjoint of

$$u_t : A_t \to A, \qquad (X \xrightarrow{g} Y) \mapsto (tX \xrightarrow{tg} t^2Y \xrightarrow{\mu Y} tY).$$

(2) Verify that the adjunction of part (1) induces—in the sense of Example 2.21 2—the given monad (t, η, μ).

Solution. For part (1) let us construct the unit and counit of the stated adjunction. The composite functor $u_t f_t$ is equal to t since η is the unit of μ. Hence the unit of the adjunction $f_t \dashv u_t$ should be a natural transformation $1 \to u_t f_t = t$. Let its component at an arbitrary object X be $\eta_X : X \to tX$; it is natural by definition. The counit should be a natural transformation from $f_t u_t : A_t \to A_t$ to the identity functor; let its component $\varepsilon_X : tX \nrightarrow X$ at an arbitrary object X be provided by the identity morphism $1 : tX \to tX$ in A. Its naturality means the commutativity of the first diagram below in A_t, for any morphism $g : X \nrightarrow Y$.

$$
\begin{array}{ccc}
f_t u_t X & \xrightarrow{f_t u_t g} & f_t u_t Y \\
{\scriptstyle \varepsilon_X}\downarrow & & \downarrow{\scriptstyle \varepsilon_Y} \\
X & \xrightarrow{\quad g \quad} & Y
\end{array}
\qquad\qquad
\begin{array}{ccc}
tX & \xrightarrow{f_t u_t g} & t^2 Y \\
{\scriptstyle tg}\downarrow & {\scriptstyle \eta_t Y}\nearrow & \downarrow{\scriptstyle \mu Y} \\
t^2 Y \xrightarrow{\mu Y} tY & = & tY
\end{array}
$$

This holds by the commutativity of the second diagram in A. The triangle identities (2.2) on the so constructed unit and counit hold by the unitality of the monad (t, η, μ):

$$(t^2 X \xrightarrow{u_t \varepsilon X} tX)\circ(tX \xrightarrow{\eta_{u_t} X} t^2 X) = tX \xrightarrow{\eta_t X} t^2 X \underset{u_t \varepsilon X}{=\!=\!=} t^2 X \xrightarrow{\mu X} tX = 1$$

$$(tX \xrightarrow{\varepsilon_{f_t} X} X)\circ(X \xrightarrow{f_t \eta X} tX) = X \xrightarrow{\eta X} tX \xrightarrow{\eta_t X} t^2 X \underset{f_t \eta X}{=\!=\!=} t^2 X \xrightarrow{\mu X} tX = X \xrightarrow{\eta X} tX .$$

Part (2) follows immediately by the explicit forms of the unit and counit of the adjunction $f_t \dashv u_t$ constructed above.

Exercise 2.26. Prove that for any adjunction $l \dashv r : \mathsf{B} \to \mathsf{A}$ inducing in the sense of Example 2.21 2 a given monad (t, η, μ) on A, there is a unique functor k from B to the Eilenberg–Moore category A^t of Definition 2.22 for which the following diagram (using the notation of Paragraph 2.24) commutes.

$$
\begin{array}{ccc}
\mathsf{A} & \xrightarrow{\ l\ } & \mathsf{B} \\
{\scriptstyle f^t}\downarrow & {\scriptstyle k}\swarrow & \downarrow{\scriptstyle r} \\
\mathsf{A}^t & \xrightarrow[\ u^t\ ]{} & \mathsf{A}
\end{array}
\tag{2.3}
$$

Solution. Let us prove first that there is at most one such functor k. From the commutativity of the lower triangle in (2.3) we know that if k exists then it should send an object B to some Eilenberg–Moore algebra consisting of the object part rB and some morphism part $\kappa_B : trB = rlrB \to rB$; and it should send a morphism b to rb. Commutativity of the upper triangle in (2.3) says in addition that for any object A of A the component $\kappa_{lA} : rlrlA \to rlA$ is equal to the component μ_A of the multiplication of the monad $t = rl$. With this information at hand we infer the commutativity of the following diagram for the counit $\varepsilon : lr \to 1$ of the adjunction $l \dashv r$ and any object B of B.

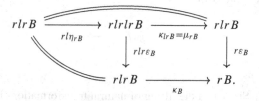

The region at the top commutes by the unitalty of the monad (t, η, μ), the triangle on the left commutes by one of the triangle identities of (2.2), and the square on the right commutes since $k\varepsilon_B = r\varepsilon_B$ is a morphism of Eilenberg–Moore algebras. This proves that κ is uniquely fixed as 1ε (which is consistent with $\kappa_{lA} = r\varepsilon_{lA} = \mu_A$).

The only possible functor

$$
k : (\, B \xrightarrow{\ b\ } B' \,) \mapsto (\, (rB, r\varepsilon_B) \xrightarrow{\ rb\ } (rB', r\varepsilon_{B'}) \,)
$$

evidently renders commutative the diagram of (2.3).

Exercise 2.29. Consider monads t on a category A, s on B and z on C; together with functors $f : \mathsf{A} \to \mathsf{B}$ and $g : \mathsf{B} \to \mathsf{C}$. Assume that they admit liftings $f^\varphi : \mathsf{A}^t \to \mathsf{B}^s$ and $g^\gamma : \mathsf{B}^s \to \mathsf{C}^z$ along some monad morphisms φ and γ. By the commutativity of the diagram

$$
\begin{array}{ccccc}
\mathsf{A}^t & \xrightarrow{\;f^\varphi\;} & \mathsf{B}^s & \xrightarrow{\;g^\gamma\;} & \mathsf{C}^z \\
{\scriptstyle u^t}\big\downarrow & & {\scriptstyle u^s}\big\downarrow & & {\scriptstyle u^z}\big\downarrow \\
\mathsf{A} & \xrightarrow[f]{} & \mathsf{B} & \xrightarrow[g]{} & \mathsf{C}
\end{array}
$$

we know that $g^\gamma f^\varphi$ is a lifting of gf. Compute the corresponding monad morphism.

Solution. By the construction of the data in part (i) from those in part (ii) of Theorem and Definition 2.27, for any t-algebra $(X, x : TX \to X)$ we have

$$
f^\varphi(X, x) = (fX, \ sfX \xrightarrow{\;\varphi X\;} ftX \xrightarrow{\;fx\;} fX \ ) \quad \text{and}
$$

$$
g^\gamma f^\varphi(X, x) = (gfX, \ zgfX \xrightarrow{\;\gamma fX\;} gsfX \xrightarrow{\;g\varphi X\;} gftX \xrightarrow{\;gfx\;} gfX \ ).
$$

Then by the inverse construction in Theorem and Definition 2.27, we can read off the desired monad morphism

$$
zgf \xrightarrow{\;\gamma 1\;} gsf \xrightarrow{\;1\varphi\;} gft.
$$

Exercise 2.32. Show that the lifting of natural transformations is compatible with the composition and the Godement product in Paragraph 2.9. More precisely, consider monads t, s and z on respective categories A, B and C. Let k, h and g be functors $\mathsf{A} \to \mathsf{B}$ all of which have liftings k^κ, h^χ and $g^\gamma : \mathsf{A}^t \to \mathsf{B}^s$ (along respective monad morphisms κ, χ and γ) and let h' and g' be functors $\mathsf{B} \to \mathsf{C}$ which have liftings $h'^{\chi'}$ and $g'^{\gamma'} : \mathsf{B}^s \to \mathsf{C}^z$ (along respective monad morphisms χ' and γ'). Verify the following claims.

(1) If both natural transformations $\omega : k \to h$ and $\vartheta : h \to g$ admit liftings $\overline{\omega} : k^\kappa \to h^\chi$ and $\overline{\vartheta} : h^\chi \to g^\gamma$ then the composite $\vartheta \circ \omega : k \to g$ also admits the lifting $\overline{\vartheta} \circ \overline{\omega} : k^\kappa \to g^\gamma$.

(2) If both natural transformations $\vartheta : h \to g$ and $\vartheta' : h' \to g'$ admit liftings $\overline{\vartheta} : h^\chi \to g^\gamma$ and $\overline{\vartheta}' : h'^{\chi'} \to g'^{\gamma'}$ then the Godement product $\vartheta'\vartheta : h'h \to g'g$ also admits the lifting $\overline{\vartheta}'\overline{\vartheta}$.

Solution. Both claims follow by the commutativity of the respective diagram of natural transformations

Left diagram:

$$
\begin{array}{ccc}
u^s k^\kappa \xrightarrow{1\overline{\omega}} u^s h^\chi \xrightarrow{1\overline{\vartheta}} u^s g^\gamma \\
\Big\| \qquad\qquad \Big\| \qquad\qquad \Big\| \\
ku^t \xrightarrow{\omega 1} hu^t \xrightarrow{\vartheta 1} gu^t
\end{array}
$$

with top arc $1(\overline{\vartheta}\circ\overline{\omega})$ and bottom arc $(\vartheta\circ\omega)1$.

Right diagram:

$$
\begin{array}{ccc}
u^z h'^{\chi'} h^\chi \xrightarrow{11\overline{\vartheta}} u^z h'^{\chi'} g^\gamma \xrightarrow{1\overline{\vartheta}'1} u^z g'^{\gamma'} g^\gamma \\
\Big\| \qquad\qquad \Big\| \qquad\qquad \Big\| \\
h' u^s h^\chi \xrightarrow{11\overline{\vartheta}} h' u^s g^\gamma \xrightarrow{\overline{\vartheta}'11} g' u^s g^\gamma \\
\Big\| \qquad\qquad \Big\| \qquad\qquad \Big\| \\
h' h u^t \xrightarrow{1\vartheta 1} h' g u^t \xrightarrow{\vartheta'11} g' g u^t.
\end{array}
$$

with top arc $1\overline{\vartheta}'\overline{\vartheta}$ and bottom arc $\vartheta'\vartheta 1$.

Exercise 3.3. Verify the commutativity of the following diagrams for the associativity and unit constraints of an arbitrary monoidal category $(A, I, \otimes)$, and arbitrary objects X, Y.

$$
\begin{array}{ccc}
(X \otimes Y) \otimes I \xrightarrow{\ \alpha_{X,Y,I}\ } X \otimes (Y \otimes I) \\
\qquad\searrow{\scriptstyle \varrho_{X\otimes Y}} \qquad \swarrow{\scriptstyle 1\otimes\varrho_Y} \\
X \otimes Y
\end{array}
$$

$$
\begin{array}{ccc}
& X \otimes Y & \\
\nearrow{\scriptstyle \lambda_X\otimes 1} && \nwarrow{\scriptstyle \lambda_{X\otimes Y}} \\
(I \otimes X) \otimes Y \xrightarrow{\ \alpha_{I,X,Y}\ } I \otimes (X \otimes Y)
\end{array}
$$

Solution. This was proven in [68].

For the first diagram, take the pentagon axiom for $Z = V = I$. Post-compose the top path with

$$
X \otimes (Y \otimes (I \otimes I)) \xrightarrow{1\otimes(1\otimes\lambda_I)} X \otimes (Y \otimes I) \xrightarrow{\alpha^{-1}_{X,Y,I}} (X \otimes Y) \otimes I \xrightarrow{\varrho_{X\otimes Y}} X \otimes Y.
$$

(8.4)

Use the naturality of α, the triangle axiom and the naturality of ϱ to see that the resulting morphism is equal to

$$
((X \otimes Y) \otimes I) \otimes I \xrightarrow{\varrho_{(X\otimes Y)}\otimes I} (X \otimes Y) \otimes I \xrightarrow{\varrho_{X\otimes Y}} X \otimes Y.
$$

Post-compose also the other path of the pentagon axiom for $Z = V = I$ with (8.4). Use now first the triangle axiom, and then the naturality of α and ϱ to see that the resulting morphism is equal to

$$((X \otimes Y) \otimes I) \otimes I \xrightarrow{\varrho_{(X \otimes Y)} \otimes I} (X \otimes Y) \otimes I \xrightarrow{\alpha_{X,Y,I}} X \otimes (Y \otimes I) \xrightarrow{1 \otimes \varrho_Y} X \otimes Y.$$

Since $\varrho_{(X \otimes Y) \otimes I}$ is an isomorphism, this proves the commutativity of the first diagram. The second diagram is handled symmetrically.

Exercise 3.4. Prove that for the left and right unit constraints λ and ϱ in any monoidal category $(\mathsf{A}, I, \otimes)$, the morphisms λ_I and $\varrho_I : I \otimes I \to I$ are equal.

Solution. This was proven in [68] too.

In the first diagram of

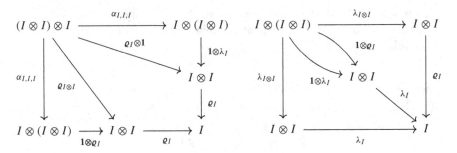

the top-right triangle commutes by the triangle axiom, the bottom-left triangle was shown to commute in Exercise 3.3 and the middle region commutes by the naturality of ϱ. Thus the paths around it are equal. Since $\alpha_{I,I,I}$ and ϱ_I are isomorphisms, this proves that the region of the second diagram bounded by the curved arrows commutes. The other regions of the second diagram commute by the naturality of λ. Since $\lambda_{I \otimes I}$ is an isomorphism, commutativity of the second diagram proves the claim.

Exercise 3.9. Consider equivalence functors $f : \mathsf{A} \to \mathsf{B}$ and $g : \mathsf{B} \to \mathsf{A}$ and a strict monoidal structure $(I, \otimes)$ on A. Construct a monoidal structure on B with respect to which f is strong (but not necessarily strict!) monoidal.

Solution. Recall from Exercise 2.19 that f is the right adjoint of g with invertible unit $\eta : 1 \to fg$ and invertible counit $\varepsilon : gf \to 1$.

As the monoidal unit of B, choose $J := fI$. For the monoidal product put

$$\circledast := (\mathsf{B} \times \mathsf{B} \xrightarrow{g \times g} \mathsf{A} \times \mathsf{A} \xrightarrow{\otimes} \mathsf{A} \xrightarrow{f} \mathsf{B}).$$

The unit constraints should be

$$J \circledast X = f(gfI \otimes gX) \xrightarrow{f(\varepsilon_I \otimes 1)} fgX \xrightarrow{\eta_X^{-1}} X$$

$$X \circledast J = f(gX \otimes gfI) \xrightarrow{f(1 \otimes \varepsilon_I)} fgX \xrightarrow{\eta_X^{-1}} X$$

and the associativity constraint should be

$$(X \circledast Y) \circledast Z \qquad\qquad\qquad\qquad X \circledast (Y \circledast Z)$$

$$f(gf(gX \otimes gY) \otimes gZ) \xrightarrow{f(\varepsilon \otimes 1)} f(gX \otimes gY \otimes gZ) \xrightarrow{f(1 \otimes \varepsilon^{-1})} f(gX \otimes gf(gY \otimes gZ))$$

for any objects X, Y, Z of B.

The triangle axiom follows applying f to the equal paths around the commutative diagram

$$gf(gX \otimes gfI) \otimes gY \xrightarrow{\varepsilon \otimes 1} gX \otimes gfI \otimes gY \xrightarrow{1 \otimes \varepsilon^{-1}} gX \otimes gf(gfI \otimes gY)$$

with vertical maps $gf(1 \otimes \varepsilon_I) \otimes 1$, $1 \otimes \varepsilon_I \otimes 1$, $1 \otimes gf(\varepsilon_I \otimes 1)$ and bottom row

$$gfgX \otimes gY \xrightarrow{g\eta_X^{-1} \otimes 1 = \varepsilon_{gX} \otimes 1} gX \otimes gY \xleftarrow{1 \otimes g\eta_Y^{-1} = 1 \otimes \varepsilon_{gY}} gX \otimes gfgY$$

for all objects X, Y of B, where the equalities of both forms of the morphisms in the bottom row follow by a triangle condition in (2.2) and both squares commute by the naturality of ε and the functoriality of $\otimes$. In order to verify the pentagon axiom we use that the first diagram of Fig. 8.1 commutes, for any objects P, Q, R, S of A, by the naturality of ε and the functoriality of $\otimes$. The pentagon axiom is obtained by taking the equal morphisms around the first diagram of Fig. 8.1 for the objects $P = gX$, $Q = gY$, $R = gZ$ and $S = gV$ for arbitrary objects X, Y, Z, V of B and applying f to them. So we equipped B with a monoidal structure.

As the strong monoidal structure of f, take the trivial nullary part $\mathbf{1} : J = fI \to fI$ and the binary part $f(\varepsilon_P \otimes \varepsilon_Q) : fP \circledast fQ = f(gfP \otimes gfQ) \to f(P \otimes Q)$ for any objects P, Q of A. Unitality holds by the commutativity of

$$f(gfI \otimes gfP) \xrightarrow{f(\varepsilon_I \otimes \varepsilon_P)} fP \xleftarrow{f(\varepsilon_P \otimes \varepsilon_I)} f(gfP \otimes gfI)$$

with vertical maps $f(\varepsilon_I \otimes 1)$, identity, $f(1 \otimes \varepsilon_I)$ and bottom row

$$fgfP \xrightarrow{\eta_{fP}^{-1} = f\varepsilon_P} fP \xleftarrow{\eta_{fP}^{-1} = f\varepsilon_P} fgfP$$

for any object P of A, where again both forms of the morphisms in the bottom row are equal by a triangle condition in (2.2). Associativity follows applying f to the second diagram of Fig. 8.1, for any objects P, Q, R of A. Commutativity of the

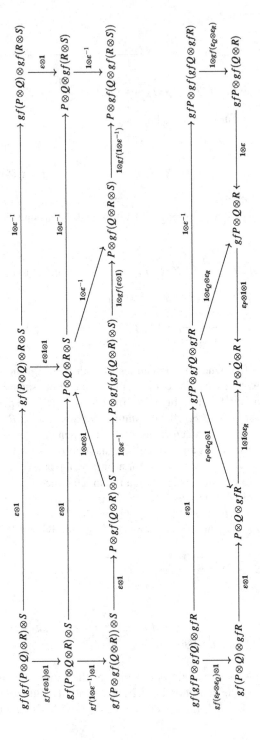

Fig. 8.1 Pentagon axiom and associativity

second diagram of Fig. 8.1 follows again by the naturality of ε and the functoriality of $\otimes$.

Exercise 3.10. Show that the composite of monoidal functors is monoidal; symmetrically, the composite of opmonoidal functors is opmonoidal.

Solution. Take functors $A \xrightarrow{f} B \xrightarrow{g} C$ with respective monoidal structures (f^0, f^2) and (g^0, g^2). Construct a monoidal structure on their composite with nullary and binary parts

$$I \xrightarrow{g^0} gI \xrightarrow{gf^0} gfI \qquad\qquad gfX \otimes gfY \xrightarrow{g^2_{fX,fY}} g(fX \otimes fY) \xrightarrow{gf^2_{X,Y}} gf(X \otimes Y).$$

By the naturality of g^2 and the associativity conditions on f^2 and g^2, the associativity diagram

$$
\begin{array}{ccccc}
gfX \otimes gfY \otimes gfZ & \xrightarrow{g^2_{fX,fY}\otimes 1} & g(fX \otimes fY) \otimes gfZ & \xrightarrow{gf^2_{X,Y}\otimes 1} & gf(X \otimes Y) \otimes gfZ \\
{\scriptstyle 1\otimes g^2_{fY,fZ}}\downarrow & & \downarrow{\scriptstyle g^2_{fX\otimes fY,fZ}} & & \downarrow{\scriptstyle g^2_{f(X\otimes Y),fZ}} \\
gfX \otimes g(fY \otimes fZ) & \xrightarrow{g^2_{fX,fY\otimes fZ}} & g(fX \otimes fY \otimes fZ) & \xrightarrow{g(f^2_{X,Y}\otimes 1)} & g(f(X \otimes Y) \otimes fZ) \\
{\scriptstyle 1\otimes gf^2_{Y,Z}}\downarrow & & \downarrow{\scriptstyle g(1\otimes f^2_{Y,Z})} & & \downarrow{\scriptstyle gf^2_{X\otimes Y,Z}} \\
gfX \otimes gf(Y \otimes Z) & \xrightarrow{g^2_{fX,f(Y\otimes Z)}} & g(fX \otimes f(Y \otimes Z)) & \xrightarrow{gf^2_{X,Y\otimes Z}} & gf(X \otimes Y \otimes Z)
\end{array}
$$

commutes. Similarly, by the naturality of g^2 and the unitality conditions on f^2 and g^2, the unitality diagram

$$
\begin{array}{ccccc}
gfX & \xrightarrow{g^0\otimes 1} & gI \otimes gfX & \xrightarrow{gf^0\otimes 1} & gfI \otimes gfX \\
{\scriptstyle 1\otimes g^0}\downarrow & {\scriptstyle g^2_{fX,I}} & \downarrow{\scriptstyle g^2_{I,fX}} & & \downarrow{\scriptstyle g^2_{fI,fX}} \\
gfX \otimes gI & \xrightarrow{} & gfX & \xrightarrow{g(f^0\otimes 1)} & g(fI \otimes fX) \\
{\scriptstyle 1\otimes gf^0}\downarrow & {\scriptstyle g(1\otimes f^0)}\downarrow & & & \downarrow{\scriptstyle gf^2_{I,X}} \\
gfX \otimes gfI & \xrightarrow{g^2_{fX,fI}} & g(fX \otimes fI) & \xrightarrow{gf^2_{X,I}} & gfX
\end{array}
$$

commutes.

Exercise 3.11. Prove that in an adjunction $l \dashv r$ between monoidal categories, there is a bijective correspondence between the monoidal structures on r and the opmonoidal structures on l.

Solution. Take an adjunction $l \dashv r : \mathsf{A} \to \mathsf{B}$ between monoidal categories, with unit η and counit ε. In terms of a monoidal structure (r^0, r^2) on r, construct an opmonoidal structure on l with nullary and binary parts

$$lI \xrightarrow{lr^0} lrI \xrightarrow{\varepsilon_I} I \qquad l(X \otimes Y) \xrightarrow{l(\eta_X \otimes \eta_Y)} l(rlX \otimes rlY) \xrightarrow{lr^2_{lX,lY}} lr(lX \otimes lY) \xrightarrow{\varepsilon_{lX \otimes lY}} lX \otimes lY$$

for all objects X, Y of B. In order to see the coassociativity of this opmonoidal structure, note that both diagrams of Fig. 8.2 commute by the naturality of ε, r^2 and η and a triangle condition in (2.2). Their left-bottom paths are equal by the coassociativity of r^2, hence their top-right paths are equal too.

Similarly, by the naturality of ε, r^2 and η and by a triangle condition in (2.2), both diagrams below commute.

Their left-bottom paths are equal to the identity morphism by the unitality of r and the other triangle condition in (2.2). Then also the top-right paths are equal to the identity morphism proving the counitality of l. Thus we associated an opmonoidal structure of l to any monoidal structure of r.

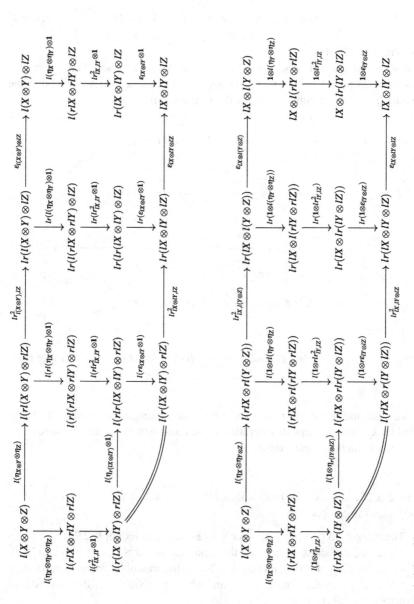

Fig. 8.2 Coassociativity of l

The inverse construction goes symmetrically. If (l^0, l^2) is an opmonoidal structure on l, then a monoidal structure on r is given by the nullary and binary parts

$$I \xrightarrow{\eta_I} rlI \xrightarrow{rl^0} rl \qquad rV \otimes rZ \xrightarrow{\eta_{rV \otimes rZ}} rl(rV \otimes rZ) \xrightarrow{rl^2_{rV,rZ}} r(lrV \otimes lrZ) \xrightarrow{r(\varepsilon_V \otimes \varepsilon_Z)} r(V \otimes Z)$$

for all objects V, Z of **A**.

It remains to see that these constructions are mutual inverses. Starting with a monoidal structure (r^0, r^2) of r and iterating both constructions we re-obtain the same data (r^0, r^2) by commutativity of the following diagrams.

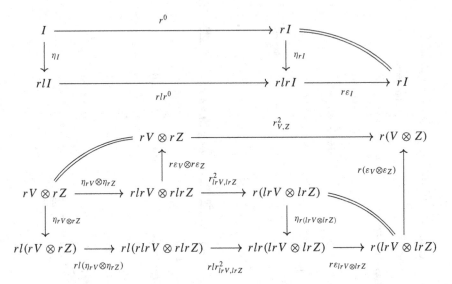

Here we made use of the naturality of η and r^2 and a triangle condition in (2.2). It is verified by symmetric steps that iterating these constructions in the opposite order we also re-obtain the original data.

Exercise 3.14. Prove the isomorphism (in the sense of Example 2.13 1) of the following categories.

- The category **alg** of algebras over a given field k in Example 2.2 8.
- The category whose objects are the monoidal functors from the monoidal singleton category $\mathbb{1}$ of Example 3.2 7 to the monoidal category **vec** of k-vector spaces in Example 3.2 2.; and whose morphisms are the monoidal natural transformations.

Symmetrically, prove that the category of coalgebras is isomorphic to the category of opmonoidal functors $\mathbb{1} \to$ **vec**.

Solution. A functor $\mathbb{1} \to$ vec is uniquely determined by the image of the single object of $\mathbb{1}$ (see Example 2.5 6); let us denote this vector space by A. The nullary part of a monoidal structure on this functor is a linear map $i : k \to A$ and the binary part is a linear map $m : A \otimes A \to A$. The associativity and unitality axioms of monoidal functors in Definition 3.5 precisely say that m is associative with the unit i.

Take now two algebras A and A' and regard them as monoidal functors $\mathbb{1} \to$ vec. A natural transformation between them consists of a single linear map $A \to A'$. The diagrams of Definition 3.12 say that it is multiplicative and preserves the unit.

The second stated isomorphism follows symmetrically.

Exercise 3.15. Consider an adjunction $l \dashv r$ and a strong monoidal structure (r^0, r^2) on r; then there is a corresponding opmonoidal structure (l^0, l^2) on l in Exercise 3.11. Regard on r the opmonoidal structure provided by the inverses of r^0 and r^2; and on the composite functors rl and lr take the opmonoidal structures in Exercise 3.10. Prove that with respect to these structures the unit $\eta : 1 \to rl$ and the counit $\varepsilon : lr \to 1$ of the adjunction are opmonoidal natural transformations.

Solution. Opmonoidality of the unit η follows by its naturality and a triangle condition in (2.2); that is, by the commutativity of the following diagrams for any objects X and Y in the domain category of l.

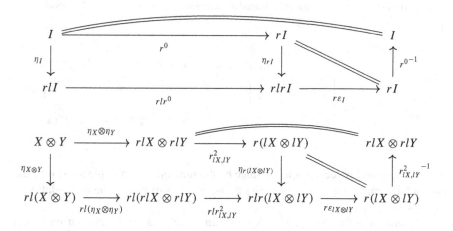

The morphisms of the bottom rows are rl^0 and $rl^2_{X,Y}$, respectively.

Opmonoidality of the counit ε follows by the naturality of ε and r^2, and by a triangle condition in (2.2) again; that is, by the commutativity of the following diagrams for any objects P and Q in the domain category of r.

$$
\begin{array}{ccccc}
lI & \xrightarrow{lr^0} & lrI & \xrightarrow{\;\;\varepsilon I\;\;} & I \\
\uparrow{\scriptstyle lr^{0-1}} & & \big\| & & \big\| \\
lrI & \xrightarrow{\hspace{5cm}} & & & I \\
& & \varepsilon I
\end{array}
$$

$$
\begin{array}{ccccc}
l(rP \otimes rQ) & \xrightarrow{l(\eta_{rP} \otimes \eta_{rQ})} & l(rlrP \otimes rlrQ) & \xrightarrow{lr^2_{lrP,lrQ}} & lr(lrP \otimes lrQ) \xrightarrow{\;\varepsilon_{lrP \otimes lrQ}\;} lrP \otimes lrQ \\
\uparrow{\scriptstyle lr^{2}_{P,Q}{}^{-1}} & & \downarrow{\scriptstyle l(r\varepsilon_P \otimes r\varepsilon_Q)} & & \downarrow{\scriptstyle lr(\varepsilon_P \otimes \varepsilon_Q)} \qquad\qquad \downarrow{\scriptstyle \varepsilon_P \otimes \varepsilon_Q} \\
lr(P \otimes Q) & & l(rP \otimes rQ) \xrightarrow[\;lr^2_{P,Q}\;]{} lr(P \otimes Q) & \xrightarrow[\;\varepsilon_{P \otimes Q}\;]{} & P \otimes Q
\end{array}
$$

The morphisms of the top rows are l^0 and $l^2_{rP,rQ}$, respectively.

The proof can also be derived from the more general results of [69], see its Section 2.1.

Exercise 3.16. Show that both the composite and the Godement product in Paragraph 2.9 of monoidal natural transformations is monoidal (regarding the monoidal structure of the composite functors in Exercise 3.10). Consequently the analogous statement holds for opmonoidal natural transformations too.

Solution. For monoidal natural transformations $f \xrightarrow{\;\varphi\;} f' \xrightarrow{\;\varphi'\;} f''$, both diagrams

$$
\begin{array}{ccc}
I == I == I \\
\downarrow{\scriptstyle f^0} \quad \downarrow{\scriptstyle f'^0} \quad \downarrow{\scriptstyle f''^0} \\
fI \xrightarrow{\varphi_I} f'I \xrightarrow{\varphi'_I} f''I
\end{array}
\qquad\qquad
\begin{array}{ccc}
fX \otimes fY & \xrightarrow{\varphi_X \otimes \varphi_Y} f'X \otimes f'Y \xrightarrow{\varphi'_X \otimes \varphi'_Y} f''X \otimes f''Y \\
\downarrow{\scriptstyle f^2_{X,Y}} & \qquad\qquad \downarrow{\scriptstyle f'^2_{X,Y}} \qquad\qquad\qquad \downarrow{\scriptstyle f''^2_{X,Y}} \\
f(X \otimes Y) & \xrightarrow[\varphi_{X \otimes Y}]{} f'(X \otimes Y) \xrightarrow[\varphi'_{X \otimes Y}]{} f''(X \otimes Y)
\end{array}
$$

commute for any objects X and Y. Since by the functoriality of $\otimes$ the top row of the second diagram is equal to $(\varphi'_X \circ \varphi_X) \otimes (\varphi'_Y \circ \varphi_Y)$, this proves the monoidality of the composite $\varphi' \circ \varphi$.

For monoidal functors $f, f' : \mathsf{A} \to \mathsf{B}$ and $g, g' : \mathsf{B} \to \mathsf{C}$, and for monoidal natural transformations $\varphi : f \to f'$ and $\gamma : g \to g'$, both diagrams

$$
\begin{array}{ccccc}
I & =\!= & I \\
{\scriptstyle g^0}\downarrow & & \downarrow{\scriptstyle g'^0} \\
gI & =\!=\!=\!=\!=\!= gI \xrightarrow{\;\gamma_I\;} & g'I \\
{\scriptstyle gf^0}\downarrow & \downarrow{\scriptstyle gf'^0} & \downarrow{\scriptstyle g'f'^0} \\
gfI \xrightarrow{g\varphi_I} & gf'I \xrightarrow{\;\gamma_{f'I}\;} & g'f'I
\end{array}
$$

$$
\begin{array}{ccccc}
gfX \otimes gfY & \xrightarrow{g\varphi_X \otimes g\varphi_Y} & gf'X \otimes gf'Y & \xrightarrow{\gamma_{f'X}\otimes\gamma_{f'Y}} & g'f'X \otimes g'f'Y \\
{\scriptstyle g^2_{fX,fY}}\downarrow & & \downarrow{\scriptstyle g^2_{f'X,f'Y}} & & \downarrow{\scriptstyle g'^2_{f'X,f'Y}} \\
g(fX \otimes fY) & \xrightarrow{g(\varphi_X\otimes\varphi_Y)} & g(f'X \otimes f'Y) & \xrightarrow{\gamma_{f'X\otimes f'Y}} & g'(f'X \otimes f'Y) \\
{\scriptstyle gf^2_{X,Y}}\downarrow & & \downarrow{\scriptstyle gf'^2_{X,Y}} & & \downarrow{\scriptstyle g'f'^2_{X,Y}} \\
gf(X \otimes Y) & \xrightarrow{g\varphi_{X\otimes Y}} & gf'(X \otimes Y) & \xrightarrow{\gamma_{f'(X\otimes Y)}} & g'f'(X \otimes Y)
\end{array}
$$

commute by the naturality of γ and g^2 and by the monoidality conditions on φ and γ. Again by the functoriality of $\otimes$ this proves that the Godement product $\gamma\varphi : gf \to g'f'$ is monoidal.

Exercise 4.7. Show that the antipode of a Hopf algebra T is an algebra homomorphism from T to the opposite algebra T^{op}. Symmetrically, show that the antipode is a coalgebra homomorphism as well from T to the opposite coalgebra.

Solution. For the comultiplication we use the Sweedler–Heynemann implicit summation index convention of Paragraph 4.2, we denote the multiplication by juxtaposition of elements, the unit by 1 and the counit by e.

Unitality of the antipode σ is checked by the computation

$$
\sigma(1) = \sigma(1)1 = \sigma(1_1)1_2 = e(1)1 = 1,
$$

where in the first equality we used the unitality of the multiplication, in the second and the fourth equalities the unitality of the comultiplication and of the counit, respectively, and one of the antipode axioms in the third equality.

For the multiplicativity of σ, associativity and coassociativity of T are used together with the following properties. In the first, third and last equalities the unitality of the multiplication and the counitality of the comultiplication, in the second, fourth and sixth equalities one of the antipode axioms, and finally the multiplicativity of the comultiplication and the counit, respectively, in the fifth and the penultimate equalities. With these steps we obtain the following equalities, for all $g, h \in T$

$$\sigma(hg) = \sigma(h_1g)e(h_2)1 = \sigma(h_1g)h_2\sigma(h_3) = \sigma(h_1g_1)h_2e(g_2)1\sigma(h_3)$$
$$= \sigma(h_1g_1)h_2g_2\sigma(g_3)\sigma(h_3) = \sigma((h_1g_1)_1)(h_1g_1)_2\sigma(g_2)\sigma(h_2)$$
$$= 1e(h_1g_1)\sigma(g_2)\sigma(h_2) = 1e(g_1)\sigma(g_2)e(h_1)\sigma(h_2) = \sigma(g)\sigma(h).$$

For the counitality of σ use in the first equality that e is the counit, its multiplicativity and unitality, respectively, in the second and last equalities, and one of the antipode axioms in the penultimate equality. This yields the following equalities for all $h \in T$

$$e\sigma(h) = e\sigma(h_1)e(h_2) = e(\sigma(h_1)h_2) = e(h)e(1) = e(h).$$

Finally, for the comultiplicativity of σ, in addition to the associativity and coassociativity of T, the following properties are used. In the first, third and last equalities the unitality of the multiplication and the counitality of the comultiplication. In the second, fourth and penultimate equalities one of the antipode axioms. In the fifth and last equalities that the comultiplication is multiplicative and unital, respectively. These steps result in the following equalities for all $h \in T$

$$\sigma(h)_1 \otimes \sigma(h)_2 = \sigma(h_1)_1 e(h_2)1 \otimes \sigma(h_1)_2 = \sigma(h_1)_1 h_2\sigma(h_3) \otimes \sigma(h_1)_2$$
$$= \sigma(h_1)_1 h_2\sigma(h_4) \otimes \sigma(h_1)_2 e(h_3)1$$
$$= \sigma(h_1)_1 h_2\sigma(h_5) \otimes \sigma(h_1)_2 h_3\sigma(h_4)$$
$$= (\sigma(h_1)h_2)_1\sigma(h_4) \otimes (\sigma(h_1)h_2)_2\sigma(h_3)$$
$$= e(h_1)1_1\sigma(h_3) \otimes 1_2\sigma(h_2) = \sigma(h_2) \otimes \sigma(h_1).$$

Exercise 5.4. Show that in any $B|B$-coring C, the image of the comultiplication is central in a suitable B-bimodule; concretely, for any $c \in C$ and $b \in B$, $c_1 \cdot (1 \otimes b) \otimes_B c_2 = c_1 \otimes_B c_2 \cdot (b \otimes 1)$.

Solution. Use the axioms of Definition 5.3 to see that for any $c \in C$ and $b \in B$, the following equalities hold

$$c \cdot (1 \otimes b) \overset{(b)}{=} (1 \otimes \epsilon((c \cdot (1 \otimes b))_2)) \cdot (c \cdot (1 \otimes b))_1 \tag{8.5}$$
$$\overset{(c)}{=} (1 \otimes \epsilon(c_2 \cdot (1 \otimes b))) \cdot c_1 \overset{(d)}{=} (1 \otimes \epsilon(c_2 \cdot (b \otimes 1))) \cdot c_1$$
$$c \cdot (b \otimes 1) \overset{(b)}{=} (\epsilon((c \cdot (b \otimes 1))_1) \otimes 1) \cdot (c \cdot (b \otimes 1))_2 \tag{8.6}$$
$$\overset{(c)}{=} (\epsilon(c_1 \cdot (b \otimes 1)) \otimes 1) \cdot c_2 \overset{(d)}{=} (\epsilon(c_1 \cdot (1 \otimes b)) \otimes 1) \cdot c_2.$$

Putting them together,

$$c_1 \cdot (1 \otimes b) \otimes_B c_2 \overset{(8.5)}{=} (1 \otimes \epsilon(c_{12} \cdot (b \otimes 1))) \cdot c_{11} \otimes_B c_2$$

$$\overset{(a)}{=} c_1 \otimes_B (\epsilon(c_{21} \cdot (b \otimes 1)) \otimes 1) \cdot c_{22} \overset{(8.6)}{=} c_1 \otimes_B c_2 \cdot (b \otimes 1).$$

Exercise 6.4. Show that the composite of separable Frobenius functors is separable Frobenius too, via the monoidal and opmonoidal structures as in Exercise 3.10.

Solution. For composable functors $A \xrightarrow{f} B \xrightarrow{f'} C$ with respective separable Frobenius structures (i^0, i^2, p^0, p^2) and (i'^0, i'^2, p'^0, p'^2), the separability condition

$$f' p^2_{X,Y} \circ p'^2_{fX,fY} \circ i'^2_{fX,fY} \circ f' i^2_{X,Y} = f' p^2_{X,Y} \circ f' i^2_{X,Y} = f'(p^2_{X,Y} \circ i^2_{X,Y}) = f'1 = 1$$

obviously holds. The first Frobenius condition of Definition 6.1 holds by the commutativity of the diagram

$$
\begin{array}{ccccc}
f'f(X \otimes Y) \otimes f'fZ & \xrightarrow{p'^2_{f(X \otimes Y),fZ}} & f'(f(X \otimes Y) \otimes fZ) & \xrightarrow{f'p^2_{X \otimes Y,Z}} & f'f(X \otimes Y \otimes Z) \\
\downarrow{\scriptstyle f'i^2_{X,Y} \otimes 1} & & \downarrow{\scriptstyle f'(i^2_{X,Y} \otimes 1)} & & \downarrow{\scriptstyle f'i^2_{X,Y \otimes Z}} \\
f'(fX \otimes fY) \otimes f'fZ & \xrightarrow{p'^2_{fX \otimes fY,fZ}} & f'(fX \otimes fY \otimes fZ) & \xrightarrow{f'(1 \otimes p^2_{Y,Z})} & f'(fX \otimes f(Y \otimes Z)) \\
\downarrow{\scriptstyle i'^2_{fX,fY} \otimes 1} & & \downarrow{\scriptstyle i'^2_{fX,fY \otimes fZ}} & & \downarrow{\scriptstyle i'^2_{fX,f(Y \otimes Z)}} \\
f'fX \otimes f'fY \otimes f'fZ & \xrightarrow{1 \otimes p'^2_{fY,fZ}} & f'fX \otimes f'(fY \otimes fZ) & \xrightarrow{1 \otimes f'p^2_{Y,Z}} & f'fX \otimes f'f(Y \otimes Z)
\end{array}
$$

where we used the naturality of i'^2 and p'^2 and the Frobenius conditions on f and f'. The other Frobenius condition of Definition 6.1 holds by a symmetric argument.

Exercise 6.5. Regard an arbitrary algebra B as a B-bimodule with actions provided by the multiplication, and regard $B \otimes B$ as a B-bimodule with left action provided by the multiplication in the first factor and right action provided by the multiplication in the second factor (that is, the B-bimodule ${}_{|}B^e$ in (5.1)). Show that any B-bimodule map $B \to B \otimes B$ with respect to these actions is in fact a coassociative (but possibly non-counital) comultiplication.

Solution. Let $\delta : B \to B \otimes B$ be a B-bimodule map; that is, a linear map rendering commutative the first diagram of (6.1). Then the following diagram commutes.

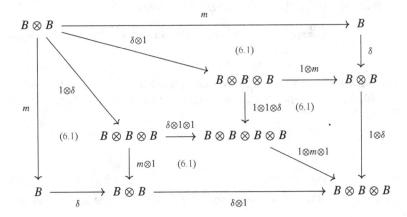

Since the multiplication occurring in the top row and the left column is an epimorphism (split by the unit applied in either factor), this proves the equality of the right column and the bottom row; that is, the coassociativity of δ.

Exercise 6.7. In terms of a separable Frobenius structure $(b \mapsto b_{(1)} \otimes b_{(2)}, \psi)$ on an algebra B, construct a separable Frobenius structure on the opposite algebra B^{op}.

Solution. The opposite comultiplication $b \mapsto b_{(2)} \otimes b_{(1)}$ is coassociative by the coassociativity of the original comultiplication and it admits the same counit ψ. Since the original comultiplication splits the original multiplication, the opposite comultiplication splits the opposite multiplication. The bilinearity condition on the original comultiplication

$$b_{(2)} \otimes b' b_{(1)} = (b'b)_{(2)} \otimes (b'b)_{(1)} = b_{(2)} b' \otimes b_{(1)}$$

for all $b, b' \in B$ can be seen as the bilinearity condition of the opposite comultiplication for the opposite multiplication.

Exercise 6.8. Show that any linear map between Frobenius algebras, which is a homomorphism of both algebras and coalgebras, is invertible.

Solution. This was proven in [91, Proposition A.3].

Let $f : B \to B'$ be a map as in the claim. The inverse is constructed as

$$B' \xrightarrow{i \otimes 1} B \otimes B' \xrightarrow{\delta \otimes 1} B \otimes B \otimes B' \xrightarrow{1 \otimes f \otimes 1} B \otimes B' \otimes B' \xrightarrow{1 \otimes m'} B \otimes B' \xrightarrow{1 \otimes \psi'} B$$

where i is the unit and δ is the comultiplication of B, and m' is the multiplication and ψ' is the counit of B'. It is indeed the inverse of f by the commutativity of the following diagrams.

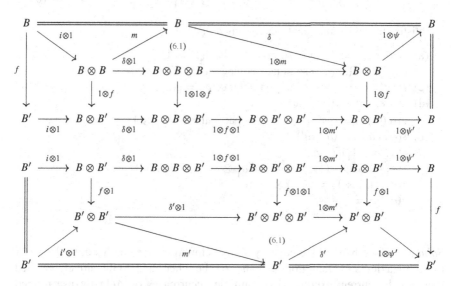

Exercise 6.9. Prove that the following categories are isomorphic (in the sense of Example 2.13 1).

- The category whose objects are the separable Frobenius algebras over a given field; and whose morphisms are those linear maps which are both algebra and coalgebra homomorphisms.
- The category whose objects are the separable Frobenius functors from the monoidal singleton category $\mathbb{1}$ of Example 3.2 7 to the monoidal category vec of vector spaces in Example 3.2 2.; and whose morphisms are those natural transformations which are both monoidal and opmonoidal.

Solution. By Exercise 3.14, a vector space B with algebra and coalgebra structures is the same as a functor $B : \mathbb{1} \to$ vec with monoidal and opmonoidal structures. Comparing Definitions 6.1 and 6.6 we immediately see that the algebra and coalgebra structures on the vector space B combine into a separable Frobenius algebra if and only if the monoidal and opmonoidal structures on the functor B combine into a separable Frobenius functor.

The bijective correspondence of morphisms in the stated categories is immediate by Exercise 3.14.

Exercise 6.16. Use the same notation as in Definition 6.15. Show that for a weak bialgebra A, the map

$$\epsilon : A \to A, \qquad a \mapsto \widehat{\epsilon}(1_{\widehat{1}}a)1_{\widehat{2}} \qquad\qquad (6.10)$$

satisfies the following identities for any $a, a' \in A$

(a) $1_{\widehat{1}} \otimes \epsilon(1_{\widehat{2}}) = 1_{\widehat{1}} \otimes 1_{\widehat{2}}$ so in particular $\epsilon(1) = 1$
(b) $\widehat{\epsilon}(aa') = \widehat{\epsilon}(a\epsilon(a'))$ so in particular $\widehat{\epsilon}\epsilon(a) = \widehat{\epsilon}(a)$
(c) $\epsilon(aa') = \epsilon(a\epsilon(a'))$ so in particular $\epsilon\epsilon(a) = \epsilon(a)$
(d) $\widehat{\Delta}(a\epsilon(a')) = a_{\widehat{1}}\epsilon(a') \otimes a_{\widehat{2}}$ and
   $\widehat{\Delta}(\epsilon(a')a) = \epsilon(a')a_{\widehat{1}} \otimes a_{\widehat{2}}$ so in particular $1_{\widehat{1}}\epsilon(a) \otimes 1_{\widehat{2}} = \epsilon(a)1_{\widehat{1}} \otimes 1_{\widehat{2}}$
(e) $\epsilon(\epsilon(a)a') = \epsilon(a)\epsilon(a')$
(f) $\epsilon(a) = \epsilon(1_{\widehat{1}})\widehat{\epsilon}(1_{\widehat{2}}a)$
(g) $\epsilon(a_{\widehat{1}}) \otimes a_{\widehat{2}} = \epsilon(1_{\widehat{1}}) \otimes 1_{\widehat{2}}a$ and $a_{\widehat{1}} \otimes \epsilon(a_{\widehat{2}}) = 1_{\widehat{1}}a \otimes 1_{\widehat{2}}$
(h) $\epsilon(a)\epsilon(1_{\widehat{1}}) \otimes 1_{\widehat{2}} = \epsilon(1_{\widehat{1}}) \otimes 1_{\widehat{2}}\epsilon(a)$
(i) $\epsilon(a_{\widehat{1}})a_{\widehat{2}} = a$.

Solution. The labels on the equality signs below refer, on one hand, to the weak bialgebra axioms in Definition 6.15 and, on the other hand, to the already verified items of the current exercise. For multiple occurrences of $\widehat{\Delta}(1)$ we use primed indices $1_{\widehat{1}} \otimes 1_{\widehat{2}}, 1_{\widehat{1}'} \otimes 1_{\widehat{2}'}$ and so on.

(a) The first equality holds by

$$1_{\widehat{1}} \otimes \epsilon(1_{\widehat{2}}) = 1_{\widehat{1}} \otimes \widehat{\epsilon}(1_{\widehat{1}'}1_{\widehat{2}})1_{\widehat{2}'} \stackrel{\text{Axiom } (b)}{=} 1_{\widehat{1}} \otimes \widehat{\epsilon}(1_{\widehat{2}})1_{\widehat{3}} = 1_{\widehat{1}} \otimes 1_{\widehat{2}}$$

and in order to obtain the second equality, apply $\widehat{\epsilon}$ to the first tensor factor.

(b) The first equality follows by

$$\widehat{\epsilon}(a\epsilon(a')) = \widehat{\epsilon}(a1_{\widehat{2}})\widehat{\epsilon}(1_{\widehat{1}}a') \stackrel{\text{Axiom } (c)}{=} \widehat{\epsilon}(aa')$$

for any $a, a' \in A$ and for the second equality put $a = 1$.

(c) The first equality follows by

$$\epsilon(aa') = \widehat{\epsilon}(1_{\widehat{1}}aa')1_{\widehat{2}} \stackrel{(b)}{=} \widehat{\epsilon}(1_{\widehat{1}}a\epsilon(a'))1_{\widehat{2}} = \epsilon(a\epsilon(a'))$$

for any $a, a' \in A$ and for the second equality put $a = 1$.

(d) The first equality follows by

$$\widehat{\Delta}(a\epsilon(a')) \quad = \quad \widehat{\epsilon}(1_{\hat{1}}a')(a1_{\hat{2}})_{\hat{1}} \otimes (a1_{\hat{2}})_{\hat{2}} \overset{\text{Axiom }(a)}{=} \widehat{\epsilon}(1_{\hat{1}}a')a_{\hat{1}}1_{\hat{2}} \otimes a_{\hat{2}}1_{\hat{3}}$$

$$\overset{\text{Axiom }(b)}{=} \widehat{\epsilon}(1_{\hat{1}}a')a_{\hat{1}}1_{\hat{1}'}1_{\hat{2}} \otimes a_{\hat{2}}1_{\hat{2}'} \overset{\text{Axiom }(a)}{=} \widehat{\epsilon}(1_{\hat{1}}a')(a1)_{\hat{1}}1_{\hat{2}} \otimes (a1)_{\hat{2}}$$

$$= \quad a_{\hat{1}}\widehat{\epsilon}(1_{\hat{1}}a')1_{\hat{2}} \otimes a_{\hat{2}} = a_{\hat{1}}\epsilon(a') \otimes a_{\hat{2}}$$

for any $a, a' \in A$ and the other equality follows symmetrically by the other condition in Axiom (b). For the final claim compare both equalities at $a = 1$.

(e) For any $a, a' \in A$,

$$\epsilon(\epsilon(a)a') = \widehat{\epsilon}(1_{\hat{1}}\epsilon(a)a')1_{\hat{2}} \overset{(b)}{=} \widehat{\epsilon}(1_{\hat{1}}\epsilon(a)\epsilon(a'))1_{\hat{2}} \overset{(d)}{=} \widehat{\epsilon}(\epsilon(a)_{\hat{1}}\epsilon(a'))\epsilon(a)_{\hat{2}}$$

$$\overset{(d)}{=} \widehat{\epsilon}([\epsilon(a)\epsilon(a')]_{\hat{1}})[\epsilon(a)\epsilon(a')]_{\hat{2}} = \epsilon(a)\epsilon(a').$$

(f) For any $a \in A$, $\epsilon(1_{\hat{1}})\widehat{\epsilon}(1_{\hat{2}}a) = \widehat{\epsilon}(1_{\hat{1}'}1_{\hat{1}})1_{\hat{2}'}\widehat{\epsilon}(1_{\hat{2}}a) \overset{\text{Axiom }(c)}{=} \widehat{\epsilon}(1_{\hat{1}'}a)1_{\hat{2}'} = \epsilon(a).$

(g) For any $a \in A$, the first equality holds by

$$\epsilon(a_{\hat{1}}) \otimes a_{\hat{2}} \overset{(f)}{=} \widehat{\epsilon}(1_{\hat{2}}a_{\hat{1}})\epsilon(1_{\hat{1}}) \otimes a_{\hat{2}} \overset{(a)}{=} \widehat{\epsilon}(\epsilon(1_{\hat{2}})a_{\hat{1}})\epsilon(1_{\hat{1}}) \otimes a_{\hat{2}}$$

$$\overset{(d)}{=} \epsilon(1_{\hat{1}}) \otimes \widehat{\epsilon}([\epsilon(1_{\hat{2}})a]_{\hat{1}})[\epsilon(1_{\hat{2}})a]_{\hat{2}} = \epsilon(1_{\hat{1}}) \otimes \epsilon(1_{\hat{2}})a \overset{(a)}{=} \epsilon(1_{\hat{1}}) \otimes 1_{\hat{2}}a$$

and the second equality holds by

$$a_{\hat{1}} \otimes \epsilon(a_{\hat{2}}) \quad = \quad a_{\hat{1}} \otimes \widehat{\epsilon}(1_{\hat{1}}a_{\hat{2}})1_{\hat{2}} \overset{\text{Axiom }(a)}{=} 1_{\hat{1}'}a_{\hat{1}}\widehat{\epsilon}(1_{\hat{1}}1_{\hat{2}'}a_{\hat{2}}) \otimes 1_{\hat{2}}$$

$$\overset{\text{Axiom }(b)}{=} 1_{\hat{1}}a_{\hat{1}}\widehat{\epsilon}(1_{\hat{2}}a_{\hat{2}}) \otimes 1_{\hat{3}} \overset{\text{Axiom }(a)}{=} (1_{\hat{1}}a)_{\hat{1}}\widehat{\epsilon}((1_{\hat{1}}a)_{\hat{2}}) \otimes 1_{\hat{2}}$$

$$= \quad 1_{\hat{1}}a \otimes 1_{\hat{2}}.$$

(h) For any $a \in A$,

$$\epsilon(a)\epsilon(1_{\hat{1}}) \otimes 1_{\hat{2}} \overset{(e)}{=} \epsilon(\epsilon(a)1_{\hat{1}}) \otimes 1_{\hat{2}} \overset{(d)}{=} \epsilon(\epsilon(a)_{\hat{1}}) \otimes \epsilon(a)_{\hat{2}} \overset{(g)}{=} \epsilon(1_{\hat{1}}) \otimes 1_{\hat{2}}\epsilon(a).$$

(i) For any $a \in A$, $\epsilon(a_{\hat{1}})a_{\hat{2}} = \widehat{\epsilon}(1_{\hat{1}}a_{\hat{1}})1_{\hat{2}}a_{\hat{2}} \overset{\text{Axiom }(a)}{=} \widehat{\epsilon}((1a)_{\hat{1}})(1a)_{\hat{2}} = \widehat{\epsilon}(a_{\hat{1}})a_{\hat{2}} = a.$

Exercise 6.21. Prove that the antipode of a weak Hopf algebra T is an algebra homomorphism from T to its opposite T^{op}.

Solution. Let us use the same notation in Definitions 6.15 and 6.20; as well as in (6.11).

Unitality of the antipode σ follows by

$$\sigma(1) = \sigma(1_{\hat{1}})\epsilon(1_{\hat{2}}) = \sigma(1_{\hat{1}})1_{\hat{2}} = \bar{\epsilon}(1) = 1.$$

The first equality follows by the first and last antipode axioms in Definition 6.20. The second equality holds by part (a) of Exercise 6.16. The third equality uses the second antipode axiom in Definition 6.20. The last equality is a symmetric variant of the last assertion in part (a) of Exercise 6.16.

Concerning its multiplicativity, for any elements g, h of T the following equalities hold

$$\sigma(hg) = \sigma((hg)_{\hat{1}})\epsilon((hg)_{\hat{2}}) = \sigma(h_{\hat{1}}g_{\hat{1}})\epsilon(h_{\hat{2}}g_{\hat{2}}) = \sigma(h_{\hat{1}}g_{\hat{1}})\epsilon(h_{\hat{2}}\epsilon(g_{\hat{2}}))$$

$$= \sigma(h_{\hat{1}}g_{\hat{1}})h_{\hat{2}}\epsilon(g_{\hat{2}})\sigma(h_{\hat{3}}) = \sigma(h_{\hat{1}}g_{\hat{1}})h_{\hat{2}}g_{\hat{2}}\sigma(g_{\hat{3}})\sigma(h_{\hat{3}})$$

$$= \sigma((h_{\hat{1}}g_{\hat{1}})_{\hat{1}})(h_{\hat{1}}g_{\hat{1}})_{\hat{2}}\sigma(g_{\hat{2}})\sigma(h_{\hat{2}}) = \bar{\epsilon}(h_{\hat{1}}g_{\hat{1}})\sigma(g_{\hat{2}})\sigma(h_{\hat{2}})$$

$$= \bar{\epsilon}(\bar{\epsilon}(h_{\hat{1}})g_{\hat{1}})\sigma(g_{\hat{2}})\sigma(h_{\hat{2}}) = \sigma(g_{\hat{1}})\bar{\epsilon}(h_{\hat{1}})g_{\hat{2}}\sigma(g_{\hat{3}})\sigma(h_{\hat{2}})$$

$$= \sigma(g_{\hat{1}})\bar{\epsilon}(h_{\hat{1}})\epsilon(g_{\hat{2}})\sigma(h_{\hat{2}}) = \sigma(g_{\hat{1}})\epsilon(g_{\hat{2}})\bar{\epsilon}(h_{\hat{1}})\sigma(h_{\hat{2}}) = \sigma(g)\sigma(h).$$

The first equality follows by the first and last antipode axioms in Definition 6.20. The second and sixth equalities are obtained by the multiplicativity of the comultiplication, that is; axiom (a) in Definition 6.15. In the third equality we used part (c) of Exercise 6.16 and in the eighth one we used its symmetric variant. In the fourth equality we applied together the first antipode axiom in Definition 6.20 with part (d) of Exercise 6.16 and in a similar way in the ninth one we applied together the second antipode axiom in Definition 6.20 with a symmetric variant of part (d) of Exercise 6.16. The fifth and tenth equalities follow by the first antipode axiom in Definition 6.20; while the seventh follows by the second antipode axiom in Definition 6.20. The penultimate equality is a consequence of the last claim in part (d) of Exercise 6.16. In the last equality we used all of the antipode axioms in Definition 6.20.

Exercise 7.5. Show that a monoid in the monoidal category **alg** of algebras over a given field k in Example 3.2 9 is precisely a commutative k-algebra.

Solution. The following is known as the *Eckmann–Hilton argument* [50].

If A is a commutative algebra then its multiplication is an algebra homomorphism $A \otimes A \to A$ and its unit defines an algebra homomorphism $k \to A$. Via these maps A is thus a monoid in **alg**.

Conversely, an arbitrary monoid in **alg** consists of an object—that is, an algebra A with multiplication $\cdot$ and unit 1—together with algebra homomorphisms $* :$

$A \otimes A \to A$ and $i : k \to A$ subject to the associativity and unitality axioms. The requirement that i preserves the unit says that it sends the multiplicative unit of k to the unit element 1 of the algebra $(A, 1, \cdot)$; that is, 1 is also the unit for the multiplication $*$. The requirement that $*$ is multiplicative says that for any $a, a', b, b' \in A$,

$$(a' * b') \cdot (a * b) = (a' \cdot a) * (b' \cdot b). \tag{8.7}$$

Putting in (8.7) b' and a equal to the common unit 1 of both multiplications $\cdot$ and $*$ we obtain the first equality in

$$a' \cdot b = a' * b \qquad\qquad b' \cdot a = a * b' \tag{8.8}$$

for any $a', b \in A$. On the other hand, putting in (8.7) a' and b equal to the common unit 1 of both multiplications $\cdot$ and $*$ we obtain the second equality in (8.8) for any $a, b' \in A$. Comparing the equalities of (8.8) we infer that both multiplications $\cdot$ and $*$ must coincide and be commutative.

Exercise 7.9. Spell out the diagrams which the structure morphisms $\xi, \xi^0, \xi_0, \xi_0^0$ of a duoidal category $(\mathsf{A}, I, \diamond, J, \blacklozenge)$ must render commutative.

Solution. The associativity diagram of Definition 3.5 for the monoidal functor $(\diamond, \xi^0, \xi)$ takes the form

$$
\begin{array}{ccc}
(X \diamond Y) \blacklozenge (X' \diamond Y') \blacklozenge (X'' \diamond Y'') & \xrightarrow{\;\xi_{X,Y,X',Y'} \blacklozenge 1\;} & ((X \blacklozenge X') \diamond (Y \blacklozenge Y')) \blacklozenge (X'' \diamond Y'') \\[2mm]
{\scriptstyle 1 \blacklozenge \xi_{X',Y',X'',Y''}} \Big\downarrow & & \Big\downarrow {\scriptstyle \xi_{X \blacklozenge X', Y \blacklozenge Y', X'', Y''}} \\[2mm]
(X \diamond Y) \blacklozenge ((X' \blacklozenge X'') \diamond (Y' \blacklozenge Y'')) & \xrightarrow{\;\xi_{X,Y,X' \blacklozenge X'',Y' \blacklozenge Y''}\;} & (X \blacklozenge X' \blacklozenge X'') \diamond (Y \blacklozenge Y' \blacklozenge Y'')
\end{array}
$$

This coincides with the compatibility condition between the opmonoidal associativity constraint of the monoidal category $(\mathsf{A}, J, \blacklozenge)$ and the binary parts of the opmonoidal functors between which it goes.

The unitality diagrams of Definition 3.5 for the monoidal functor $(\diamond, \xi^0, \xi)$ take the form

$$
\begin{array}{ccc}
X \diamond Y & \xrightarrow{\;\xi^0 \blacklozenge 1\;} & (J \diamond J) \blacklozenge (X \diamond Y) \\[2mm]
{\scriptstyle 1 \blacklozenge \xi^0} \Big\downarrow & \diagdown\diagdown & \Big\downarrow {\scriptstyle \xi_{J,J,X,Y}} \\[2mm]
(X \diamond Y) \blacklozenge (J \diamond J) & \xrightarrow{\;\xi_{X,Y,J,J}\;} & X \diamond Y
\end{array}
\tag{8.9}
$$

Its triangles coincide with the compatibility conditions between the opmonoidal left and right unit constraints of the monoidal category $(\mathbf{A}, J, \blacklozenge)$ and the binary parts of the opmonoidal functors between which they go.

The associativity condition on the monoidal functor (I, ξ_0^0, ξ_0) takes the form

$$
\begin{array}{ccc}
I \blacklozenge I \blacklozenge I & \xrightarrow{\;\xi_0 \blacklozenge 1\;} & I \blacklozenge I \\
{\scriptstyle 1 \blacklozenge \xi_0}\Big\downarrow & & \Big\downarrow{\scriptstyle \xi_0} \\
I \blacklozenge I & \xrightarrow[\;\xi_0\;]{} & I
\end{array}
$$

This coincides with the compatibility condition between the opmonoidal associativity constraint of the monoidal category $(\mathbf{A}, J, \blacklozenge)$ and the nullary parts of the opmonoidal functors between which it goes.

The unitality conditions on the monoidal functor (I, ξ_0^0, ξ_0) take the form

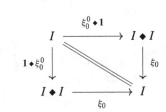

Its triangles coincide with the compatibility conditions between the opmonoidal left and right unit constraints of the monoidal category $(\mathbf{A}, J, \blacklozenge)$ and the nullary parts of the opmonoidal functors between which they go.

Next we draw the compatibility diagram between the monoidal associativity constraint of the monoidal category $(\mathbf{A}, I, \diamond)$ and the binary parts of the monoidal functors between which it goes:

$$
\begin{array}{ccc}
(X \diamond Y \diamond Z) \blacklozenge (X' \diamond Y' \diamond Z') & \xrightarrow{\;\xi_{X \diamond Y, Z, X' \diamond Y', Z'}\;} & ((X \diamond Y) \blacklozenge (X' \diamond Y')) \diamond (Z \blacklozenge Z') \\
{\scriptstyle \xi_{X, Y \diamond Z, X', Y' \diamond Z'}}\Big\downarrow & & \Big\downarrow{\scriptstyle \xi_{X, Y, X', Y'} \diamond 1} \\
(X \blacklozenge X') \diamond ((Y \diamond Z) \blacklozenge (Y' \diamond Z')) & \xrightarrow[\;1 \diamond \xi_{Y, Z, Y', Z'}\;]{} & (X \blacklozenge X') \diamond (Y \blacklozenge Y') \diamond (Z \blacklozenge Z')
\end{array}
$$

This agrees with the coassociativity condition on the opmonoidal functor $(\blacklozenge, \xi_0, \xi)$.

Next the compatibility diagram between the monoidal associativity constraint of the monoidal category $(\mathbf{A}, I, \diamond)$ and the nullary parts of the monoidal functors between which it goes:

$$
\begin{array}{ccc}
J & \xrightarrow{\ \xi^0\ } & J \bullet J \\[2pt]
{\scriptstyle \xi^0}\big\downarrow & & \big\downarrow{\scriptstyle \xi^0 \bullet 1} \\[2pt]
J \bullet J & \xrightarrow[\ 1 \bullet \xi^0\ ]{} & J \bullet J \bullet J
\end{array}
$$

This agrees with the coassociativity condition on the opmonoidal functor (J, ξ_0^0, ξ^0).

The compatibility diagrams between the monoidal left and right unit constraints of the monoidal category $(\mathbf{A}, I, \diamond)$ and the binary parts of the monoidal functors between which they go, are the triangles of

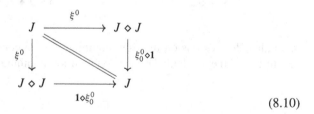

Together they coincide with the counitality conditions on the opmonoidal functor $(\bullet, \xi_0, \xi)$.

Finally, the compatibility diagrams between the monoidal left and right unit constraints of the monoidal category $(\mathbf{A}, I, \diamond)$ and the nullary parts of the monoidal functors between which they go, are the triangles of

$$
\begin{array}{ccc}
J & \xrightarrow{\ \xi^0\ } & J \diamond J \\[2pt]
{\scriptstyle \xi^0}\big\downarrow & \searrow & \big\downarrow{\scriptstyle \xi_0^0 \diamond 1} \\[2pt]
J \diamond J & \xrightarrow[\ 1 \diamond \xi_0^0\ ]{} & J
\end{array}
\tag{8.10}
$$

Together they coincide with the counitality conditions on the opmonoidal functor (J, ξ_0^0, ξ^0).

Exercise 7.10. Show that in Definition 7.8 the morphism $\xi_0^0 : J \to I$ is redundant; (omitting the unit constraints) it can be written either as

(continued)

$$J = (I \diamond J) \bullet (J \diamond I) \xrightarrow{\xi} (I \bullet J) \diamond (J \bullet I) = I \qquad \text{or}$$

$$J = (J \diamond I) \bullet (I \diamond J) \xrightarrow{\xi} (J \bullet I) \diamond (I \bullet J) = I.$$

Solution. This can be found in [3, Proposition 6.9]. The claim follows by the commutativity of the symmetric diagrams

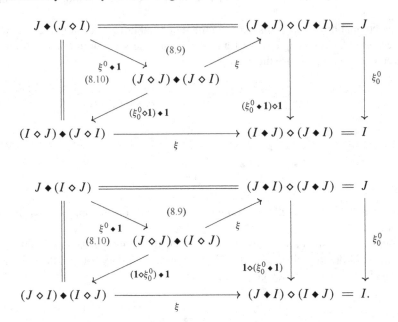

The unlabelled regions on the right commute by the coherence of both monoidal structures and the unlabelled regions on their left commute by the naturality of ξ.

Correction to: Hopf Algebras and Their Generalizations from a Category Theoretical Point of View

Correction to:
G. Böhm, *Hopf Algebras and Their Generalizations from a Category Theoretical Point of View*, **Lecture Notes in Mathematics,**
https://doi.org/10.1007/978-3-319-98137-6

The book was inadvertently published with an incomplete affiliation for Prof. Gabriella Böhm, the corrected affiliation must read 'Wigner Research Centre for Physics, Hungarian Academy of Sciences, Budapest, Hungary' instead of 'Wigner Research Centre for Physics, Budapest, Hungary'. The affiliation has been updated in the book.

The updated online version of the book can be found at
https://doi.org/10.1007/978-3-319-98137-6

© Springer Nature Switzerland AG 2019
G. Böhm, *Hopf Algebras and Their Generalizations from a Category
Theoretical Point of View*, Lecture Notes in Mathematics 2226,
https://doi.org/10.1007/978-3-319-98137-6_9

References

1. Abe, E.: Hopf Algebras. Cambridge University Press, Cambridge (1977)
2. Aguiar, M., Chase, S.U.: Generalized Hopf modules for bimonads. Theory Appl. Categ. **27**(13), 263–326 (2013)
3. Aguiar, M., Mahajan, S.: Monoidal Functors, Species and Hopf Algebras. CRM Monograph Series, vol. 29. American Mathematical Society, Providence (2010)
4. Andruskiewitsch, N., Ferrer Santos, W.: The beginnings of the theory of Hopf algebras. Acta Appl. Math. **108**, 3–17 (2009)
5. Ardizzoni, A., Böhm, G., Menini, C.: A Schneider type theorem for Hopf algebroids. J. Algebra **318**(1), 225–269 (2007). Corrigendum: J. Algebra **321**(6), 1786–1796 (2009)
6. Bálint, I., Szlachányi, K.: Finitary Galois extensions over noncommutative bases. J. Algebra **296**(2), 520–560 (2006)
7. Balteanu, C., Fiedorowicz, Z., Schwänzl, R., Vogt, R.: Iterated monoidal categories. Adv. Math. **176**(2), 277–349 (2003)
8. Batanin, M., Markl, M.: Centers and homotopy centers in enriched monoidal categories. Adv. Math. **230**(4–6), 1811–1858 (2012)
9. Batista, E., Caenepeel, S., Vercruysse, J.: Hopf categories. Algebra Represent. Theory **19**(5), 1173–1216 (2016)
10. Böhm, G.: Weak Hopf algebras and their application to spin models. PhD thesis, Eötvös University Budapest (1997)
11. Böhm, G.: Galois theory for Hopf algebroids. Ann. Univ. Ferrara Sez. VII (N.S.) **51**, 233–262 (2005)
12. Böhm, G.: Integral theory for Hopf algebroids. Algebra Represent. Theory **8**(4), 563–599 (2005). Erratum: Algebra Represent. Theory **13**(6), 755 (2010)
13. Böhm, G.: Hopf algebroids. In: Hazewinkel, M. (ed.) Handbook of Algebra, vol. 6, pp. 173–236. Elsevier, New York (2009)
14. Böhm, G.: Hopf polyads, Hopf categories and Hopf group monoids viewed as Hopf monads. Theory Appl. Categ. **32**(37), 1229–1257 (2017)
15. Böhm, G., Brzeziński, T.: Cleft extensions of Hopf algebroids. Appl. Categ. Struct. **14**(5–6), 431–469 (2006). Corrigendum: Appl. Categ. Struct. **17**(6), 613–620 (2009)
16. Böhm, G., Lack, S.: Hopf comonads on naturally Frobenius map-monoidales. J. Pure Appl. Algebra **220**(6), 2177–2213 (2016)
17. Böhm, G., Ştefan, D.: (Co)cyclic (co)homology of bialgebroids: an approach via (co)monads. Commun. Math. Phys. **282**(1), 239–286 (2008)

© Springer Nature Switzerland AG 2018
G. Böhm, *Hopf Algebras and Their Generalizations from a Category Theoretical Point of View*, Lecture Notes in Mathematics 2226,
https://doi.org/10.1007/978-3-319-98137-6

18. Böhm, G., Szlachányi, K.: A coassociative C^*-quantum group with nonintegral dimensions. Lett. Math. Phys. **38**(4), 437–456 (1996)
19. Böhm, G., Szlachányi, K.: Weak Hopf algebras II: representation theory, dimensions, and the Markov trace. J. Algebra **233**(1), 156–212 (2000)
20. Böhm, G., Szlachányi, K.: Hopf algebroids with bijective antipodes: axioms, integrals, and duals. J. Algebra **274**(2), 708–750 (2004)
21. Böhm, G., Szlachányi, K.: Hopf algebroid symmetry of abstract Frobenius extensions of depth 2. Commun. Algebra **32**(11), 4433–4464 (2004)
22. Böhm, G., Nill, F., Szlachányi, K.: Weak Hopf algebras: I. Integral theory and C^*-structure. J. Algebra **221**(2), 385–438 (1999)
23. Böhm, G., Lack, S., Street, R.: Weak bimonads and weak Hopf monads. J. Algebra **328**(1), 1–30 (2011)
24. Böhm, G., Chen, Y., Zhang, L.: On Hopf monoids in duoidal categories. J. Algebra **394**, 139–172 (2013)
25. Böhm, G., Gómez-Torrecillas, J., López-Centella, E.: On the category of weak bialgebras. J. Algebra **399**(1), 801–844 (2014)
26. Booker, T., Street, R.: Torsors, herds and flocks. J. Algebra **330**(1), 346–374 (2011)
27. Booker, T., Street, R.: Tannaka duality and convolution for duoidal categories. Theory Appl. Categ. **28**(6), 166–205 (2013)
28. Borel, A.: Sur la cohomologie des espaces fibres principaux et des espaces homogenes des groupes de Lie compacts. Ann. Math. **87**, 115–207 (1953)
29. Bruguières, A., Virelizier, A.: Hopf monads. Adv. Math. **215**(2), 679–733 (2007)
30. Bruguières, A., Lack, S., Virelizier, A.: Hopf monads on monoidal categories. Adv. Math. **227**(2), 745–800 (2011)
31. Brzeziński, T., Militaru, G.: Bialgebroids, $\times_A$-bialgebras and duality. J. Algebra **251**(1), 279–294 (2002)
32. Caenepeel, S., De Lombaerde, M.: A categorical approach to Turaev's Hopf group-coalgebras. Commun. Algebra **34**, 2631–2657 (2006)
33. Cartier, P.: Dualité de Tannaka des groupes et des algèbres de Lie. C. R. Acad. Sci. Paris **242**, 322–325 (1956)
34. Cartier, P.: A primer on Hopf algebras. In: Cartier, P., Moussa, P., Julia, B., Vanhove, P. (eds.) Frontiers in Number Theory, Physics, and Geometry II, pp. 537–615. Springer, Berlin (2007)
35. Chen, Y., Böhm, G.: Weak bimonoids in duoidal categories. J. Pure Appl. Algebra **218**(12), 2240–2273 (2014)
36. Chikhladze, D.: The Tannaka representation theorem for separable Frobenius functors. Algebra Represent. Theory **15**(6), 1205–1213 (2012)
37. Chikhladze, D., Lack, S., Street, R.: Hopf monoidal comonads. Theory Appl. Categ. **24**(19), 554–563 (2010)
38. Connes, A., Kreimer, D.: Hopf algebras, renormalization and noncommutative geometry. Commun. Math. Phys. **199**, 203–242 (1998)
39. Connes, A., Moscovici, H.: Rankin–Cohen brackets and the Hopf algebra of transverse geometry. Mosc. Math. J. **4**, 111–130 (2004)
40. Day, B., Street, R.: Monoidal bicategories and Hopf algebroids. Adv. Math. **129**(1), 99–157 (1997)
41. Day, B., Street, R.: Quantum category, star autonomy and quantum groupoids. In: Janelidze, G., Pareigis, B., Tholen, W. (eds.) Galois Theory, Hopf Algebras, and Semiabelian Categories. Fields Institute Communications, vol. 43, pp. 187–225. American Mathematical Society, Providence (2004)
42. Deligne, P.: Catégories tannakiennes. In: Cartier, P., Katz, N.M., Manin, Y.I., Illusie, L., Laumon, G., Ribet, K.A. (eds.) Grothendieck Festschrift, Volume II. Progress in Mathematics, vol. 87, pp. 111–195. Birkhäuser, Boston (1990)
43. Deligne, P., Milne, J.S.: Tannakian categories. In: Deligne, P., Milne, J.S., Ogus, A., Shih, K.-y. (eds.) Hodge Cycles, Motives, and Shimura Varieties. Lecture Notes in Mathematics, vol. 900, pp. 101–228. Springer, Berlin (1982)

44. Demazure, M., Gabriel, P.: Introduction to Algebraic Geometry and Algebraic Groups. North Holland, Amsterdam (1980)
45. Demazure, M., Grothendieck, A. (eds.): Séminaire de Géomètrie Algébrique du Bois Marie 1962-64 (SGA 3) I. Schémas en groupes. Lecture Notes in Mathematics, vol. 151. Springer, Berlin (1970)
46. Dieudonné, J.: Groupes de Lie et hyperalgèbres de Lie sur un corps de caractéristique $p > 0$. Comment. Math. Helv. **28**, 87–117 (1954)
47. Dijkgraaf, R., Pasquier, V., Roche, P.: Quasi-Hopf algebras, group cohomology and orbifold models. Nucl. Phys. B Proc. Suppl. **18B**, 60–72 (1990)
48. Doplicher, S., Roberts, J.E.: A new duality theory for compact groups. Invent. Math. **98**(1), 157–218 (1989)
49. Drinfel'd, V.: Квазихопфовы алгебры. Algebra i Analiz **1**(6), 114–148 (1989). English translation: Quasi-Hopf algebras. Leningrad Math. J. **1**(6), 1419–1457 (1990)
50. Eckmann, B., Hilton, P.J.: Group-like structures in general categories. I. Multiplications and comultiplications. Math. Ann. **145**(3), 227–255 (1962)
51. Eilenberg, S., Kelly, G.M.: Closed categories. In: Eilenberg, S., Harrison, D.K., Mac Lane, S., Röhrl, H. (eds.) Proceedings of the Conference of Categorical Algebra (La Jolla 1965), pp. 421–562. Springer, Berlin (1966)
52. Enock, M.: Inclusions of von Neumann algebras and quantum groupoïds II. J. Funct. Anal. **178**(1), 156–225 (2000)
53. Enock, M., Vallin, J.-M.: Inclusions of von Neumann algebras and quantum groupoïds. J. Funct. Anal. **172**(2), 249–300 (2000)
54. Etingof, P., Nikshych, D.: Dynamical quantum groups at roots of 1. Duke Math. J. **108**, 135–168 (2001)
55. Figueroa, H., Garcia-Bondía, J.M.: Combinatorial Hopf algebras in quantum field theory I. Rev. Math. Phys. **17**(8), 881–976 (2005)
56. Grandis, M., Paré, R.: Intercategories. Theory Appl. Categ. **30**(38), 1215–1255 (2015)
57. Grandis, M., Paré, R.: Intercategories: a framework for three-dimensional category theory. J. Pure Appl. Algebra **221**(5), 999–1054 (2017)
58. Halpern, E.: Twisted Polynomial Hyperalgebras. Memoirs of the American Mathematical Society, vol. 29. American Mathematical Society, Providence (1958)
59. Hochschild, G., Mostow, G.D.: Representations and representative functions of Lie groups. Ann. Math. Sec. Ser. **66**, 495–542 (1957)
60. Hopf, H.: Über die Topologie der Gruppen-Mannifaltigkeiten und ihrer Verallgemeinerungen. Ann. Math. **42**, 22–52 (1941)
61. Joyal, A., Street, R.: An introduction to Tannaka duality and quantum groups. In: Carboni, A., Pedicchio, M.C., Rosolini, G. (eds.) Category Theory. Lecture Notes in Mathematics, vol. 1488, pp. 411–492. Springer, Berlin (1991)
62. Joyal, A., Street, R.: Braided tensor categories. Adv. Math. **102**(1), 20–78 (1993)
63. Kac, G.I.: A generalization of the principle of duality for groups. Dokl. Akad. Nauk SSSR **138**, 275–278 (1961)
64. Kadison, L.: Galois theory for bialgebroids, depth two and normal Hopf subalgebras. Ann. Univ. Ferrara Sez. VII (NS) **51**, 209–231 (2005)
65. Kadison, R.V., Ringrose, J.R.: Fundamentals of the Theory of Operator Algebras, vol. I. Academic, Cambridge (1983)
66. Kadison, L., Szlachányi, K.: Bialgebroid actions on depth two extensions and duality. Adv. Math. **179**(1), 75–121 (2003)
67. Kassel, C.: Quantum Groups. Graduate Texts in Mathematics, vol. 155. Springer, Berlin (1995)
68. Kelly, G.M.: On MacLane's conditions for coherence of natural associativities, commutativities, etc. J. Algebra **1**(4), 397–402 (1964)
69. Kelly, G.M.: Doctrinal adjunction. In: Kelly, G.M. (ed.) Category Seminar Sydney 1972/1973. Lecture Notes in Mathematics, vol. 420, pp. 257–280. Springer, Berlin (1974)

70. Kostant, B.: Groups over $\mathbb{Z}$. In: Borel, A., Mostow, G.D. (eds.) Algebraic Groups and Discontinuous Subgroups. Proceedings of Symposia in Pure Mathematics, vol. 9, pp. 90–98. American Mathematical Society, Providence (1966)

71. Krähmer, U., Rovi, A.: A Lie–Rinehart algebra with no antipode. Commun. Algebra **43**(10), 4049–4053 (2015)

72. Longo, R.: A duality for Hopf algebras and for subfactors. Commun. Math. Phys. **159**, 133–150 (1994)

73. Lu, J.-H.: Hopf algebroids and quantum groupoids. Int. J. Math. **7**(1), 47–70 (1996)

74. Mac Lane, S.: Categories for the Working Mathematician. Springer, Berlin (1978)

75. Mack, G., Schomerus, V.: Quasi Hopf quantum symmetry in quantum theory. Nucl. Phys. B **370**(1), 185–230 (1992)

76. Majid, S.: Tannaka-Krein theorem for quasi-Hopf algebras and other results. Contemp. Math. **134**, 219–232 (1992)

77. Majid, S.: Foundations of Quantum Group Theory. Cambridge University Press, Cambridge (2000)

78. Manin, Y.I.: Quantum groups and non-commutative geometry. Centre de Recherches Mathématiques, Université de Montreal (1988)

79. McCrudden, P.: Opmonoidal monads. Theory Appl. Categ. **10**(19), 469–485 (2002)

80. McCurdy, M.B.: Graphical methods for Tannaka duality of weak bialgebras and weak Hopf algebras. Theory Appl. Categ. **26**(9), 233–280 (2012)

81. Mesablishvili, B., Wisbauer, R.: Bimonads and Hopf monads on categories. J. K-Theory **7**(2), 349–388 (2011)

82. Milnor, J.W., Moore, J.C.: On the structure of Hopf algebras. Ann. Math. Sec. Ser. **81**(2), 211–264 (1965)

83. Moerdijk, I.: Monads on tensor categories. J. Pure Appl. Algebra **168**(2–3), 189–208 (2002)

84. Montgomery, S.: Hopf Algebras and Their Actions on Rings. CBMS Lecture Notes, vol. 82. American Mathematical Society, Providence (1993)

85. Nikshych, D., Vainerman, L.: A characterization of depth 2 subfactors of II_1 factors. J. Funct. Anal. **171**(10), 278–307 (2000)

86. Nikshych, D., Vainerman, L.: Finite quantum groupoids and their applications. In: Montgomery, S., Schneider, H.-J. (eds.) New Directions in Hopf Algebras. Mathematical Sciences Research Institute Publications, vol. 43, pp. 211–262. Cambridge University Press, Cambridge (2002)

87. Nikshych, D., Turaev, V., Vainerman, L.: Invariants of knots and 3-manifolds from quantum groupoids. Topol. Appl. **127**, 91–123 (2003)

88. Nill, F.: Axioms for weak bialgebras. https://arxiv.org/abs/math/9805104 (Preprint)

89. Nill, F., Szlachányi, K.: Quantum chains of Hopf algebras with quantum double cosymmetry. Commun. Math. Phys. **187**(1), 159–200 (1997)

90. Nill, F., Szlachányi, K., Wiesbrock, H.-W.: Weak Hopf algebras and reducible Jones inclusions of depth 2. I: from crossed products to Jones towers. https://arxiv.org/abs/math/9806130 (Preprint)

91. Pastro, C., Street, R.: Weak Hopf monoids in braided monoidal categories. Algebra Number Theory **3**(2), 149–207 (2009)

92. Petkova, V.B., Zuber, J.-B.: The many faces of Ocneanu cells. Nucl. Phys. B **603**(3), 449–496 (2001)

93. Pfeiffer, H.: Tannaka-Kreın reconstruction and a characterization of modular tensor categories. J. Algebra **321**(12), 3714–3763 (2009)

94. Phùng, H.H.: Tannaka–Krein duality for Hopf algebroids. Israel J. Math. **167**(1), 193–225 (2008)

95. Ravenel, D.C.: Complex Cobordism and Stable Homotopy Groups of Spheres. Pure and Applied Mathematics, vol. 121. Academic, Cambridge (1986)

96. Rota, G.-C.: Hopf algebra methods in combinatorics. In: Bermond, J.-C. (ed.) Problèmes combinatoires et théorie des graphes. Colloques Internatinaux CNRS, vol. 260, pp. 363–365. CNRS, Paris (1978)

97. Ruelle, D.: Statistical Mechanics, Rigorous Results. W.A. Benjamin, New York (1969)
98. Saavedra Rivano, N.: Catègories Tannakiennes. Lecture Notes in Mathematics, vol. 265. Springer, Berlin (1972)
99. Schäppi, D.: The Formal Theory of Tannaka Duality. Astérisque Series, vol. 357. American Mathematical Society, Providence (2013)
100. Schauenburg, P.: Tannaka Duality for Arbitrary Hopf Algebras. Algebra Berichte, vol. 66. R. Fischer, Munich (1992)
101. Schauenburg, P.: Bialgebras over noncommutative rings, and a structure theorem for Hopf bimodules. Appl. Categ. Struct. **6**(2), 193–222 (1998)
102. Schauenburg, P.: Duals and doubles of quantum groupoids. In: Andruskiewitsch, N., Ferrer Santos, W.R., Schneider, H.-J. (eds.) New Trends in Hopf Algebra Theory. Contemporary Mathematics, vol. 267, pp. 273–300. American Mathematical Society, Providence (2000)
103. Schauenburg, P.: Morita base change in quantum groupoids. In: Vainerman, L. (ed.) Locally Compact Quantum Groups and Groupoids. IRMA Lectures in Mathematics and Theoretical Physics, vol. 2, pp. 79–103. De Gruyter, Berlin (2003)
104. Schauenburg, P.: Weak Hopf algebras and quantum groupoids. In: Hajac, P.M., Pusz, W. (eds.) Noncommutative Geometry and Quantum Groups. Banach Center Publications, vol. 61, pp. 171–188. Polish Academy of Sciences, Warsaw (2003)
105. Street, R.: The formal theory of monads. J. Pure Appl. Algebra **2**(2), 149–168 (1972)
106. Street, R.: Quantum Groups. A Path to Current Algebra. Cambridge University Press, Cambridge (2007)
107. Street, R.: Monoidal categories in, and linking, geometry and algebra. Bull. Belg. Math. Soc. Simon Stevin **19**(5), 769–821 (2012)
108. Sweedler, M.E.: Hopf Algebras. W.A. Benjamin, New York (1969)
109. Szlachányi, K.: Finite quantum groupoids and inclusions of finite type. In: Longo, R. (ed.) Mathematical Physics in Mathematics and Physics: Quantum and Operator Algebraic Aspects. Fields Institute Communications, vol. 30, pp. 393–407. American Mathematical Society, Providence (2001)
110. Szlachányi, K.: The monoidal Eilenberg–Moore construction and bialgebroids. J. Pure Appl. Algebra **182**(2–3), 287–315 (2003)
111. Szlachányi, K.: Galois actions by finite quantum groupoids. In: Vainerman, L. (ed.) Locally Compact Quantum Groups and Groupoids. IRMA Lectures in Mathematics and Theoretical Physics, vol. 2, pp. 105–126. De Gruyter, Berlin (2003)
112. Szlachányi, K.: Adjointable monoidal functors and quantum groupoids. In: Caenepeel, S., Van Oystaeyen, F. (eds.) Hopf Algebras in Noncommutative Geometry and Physics. Lecture Notes in Pure and Applied Mathematics, vol. 239, pp. 291–307. Marcel Dekker, New York (2005)
113. Szlachányi, K.: Fiber functors, monoidal sites and Tannaka duality for bialgebroids. https://arxiv.org/abs/0907.1578 (Preprint)
114. Szymański, W.: Finite index subfactors and Hopf algebra crossed products. Proc. Am. Math. Soc. **120**, 519–528 (1994)
115. Takeuchi, M.: Groups of algebras over $A \otimes \overline{A}$. J. Math. Soc. Jpn. **29**(3), 459–492 (1977)
116. Takeuchi, M.: $\sqrt{}$Morita theory—formal ring laws and monoidal equivalences of categories of bimodules. J. Math. Soc. Jpn. **39**(2), 301–336 (1987)
117. Turaev, V.G.: Homotopy field theory in dimension 3 and crossed group-categories. https://arxiv.org/abs/math/0005291 (Preprint)
118. Ulbrich, K.-H.: On Hopf algebras and rigid monoidal categories. Israel J. Math. **72**(1–2), 252–256 (1990)
119. Vecsernyés, P.: On the quantum symmetry of the chiral Ising model. Nucl. Phys. B **415**(3), 557–588 (1994)
120. Vecsernyés, P.: Larson–Sweedler theorem and the role of grouplike elements in weak Hopf algebras. J. Algebra **270**(2), 471–520 (2003)
121. Xu, P.: Quantum groupoids. Commun. Math. Phys. **216**(3), 539–581 (2001)

Index

© Springer Nature Switzerland AG 2018
G. Böhm, *Hopf Algebras and Their Generalizations from a Category
Theoretical Point of View*, Lecture Notes in Mathematics 2226,
https://doi.org/10.1007/978-3-319-98137-6

LECTURE NOTES IN MATHEMATICS  Springer

Editors in Chief: J.-M. Morel, B. Teissier;

Editorial Policy

1. Lecture Notes aim to report new developments in all areas of mathematics and their applications – quickly, informally and at a high level. Mathematical texts analysing new developments in modelling and numerical simulation are welcome.

   Manuscripts should be reasonably self-contained and rounded off. Thus they may, and often will, present not only results of the author but also related work by other people. They may be based on specialised lecture courses. Furthermore, the manuscripts should provide sufficient motivation, examples and applications. This clearly distinguishes Lecture Notes from journal articles or technical reports which normally are very concise. Articles intended for a journal but too long to be accepted by most journals, usually do not have this "lecture notes" character. For similar reasons it is unusual for doctoral theses to be accepted for the Lecture Notes series, though habilitation theses may be appropriate.

2. Besides monographs, multi-author manuscripts resulting from SUMMER SCHOOLS or similar INTENSIVE COURSES are welcome, provided their objective was held to present an active mathematical topic to an audience at the beginning or intermediate graduate level (a list of participants should be provided).

   The resulting manuscript should not be just a collection of course notes, but should require advance planning and coordination among the main lecturers. The subject matter should dictate the structure of the book. This structure should be motivated and explained in a scientific introduction, and the notation, references, index and formulation of results should be, if possible, unified by the editors. Each contribution should have an abstract and an introduction referring to the other contributions. In other words, more preparatory work must go into a multi-authored volume than simply assembling a disparate collection of papers, communicated at the event.

3. Manuscripts should be submitted either online at www.editorialmanager.com/lnm to Springer's mathematics editorial in Heidelberg, or electronically to one of the series editors. Authors should be aware that incomplete or insufficiently close-to-final manuscripts almost always result in longer refereeing times and nevertheless unclear referees' recommendations, making further refereeing of a final draft necessary. The strict minimum amount of material that will be considered should include a detailed outline describing the planned contents of each chapter, a bibliography and several sample chapters. Parallel submission of a manuscript to another publisher while under consideration for LNM is not acceptable and can lead to rejection.

4. In general, **monographs** will be sent out to at least 2 external referees for evaluation.

   A final decision to publish can be made only on the basis of the complete manuscript, however a refereeing process leading to a preliminary decision can be based on a pre-final or incomplete manuscript.

   Volume Editors of **multi-author works** are expected to arrange for the refereeing, to the usual scientific standards, of the individual contributions. If the resulting reports can be forwarded to the LNM Editorial Board, this is very helpful. If no reports are forwarded or if other questions remain unclear in respect of homogeneity etc, the series editors may wish to consult external referees for an overall evaluation of the volume.

5. Manuscripts should in general be submitted in English. Final manuscripts should contain at least 100 pages of mathematical text and should always include

   – a table of contents;
   – an informative introduction, with adequate motivation and perhaps some historical remarks: it should be accessible to a reader not intimately familiar with the topic treated;
   – a subject index: as a rule this is genuinely helpful for the reader.
   – For evaluation purposes, manuscripts should be submitted as pdf files.

6. Careful preparation of the manuscripts will help keep production time short besides ensuring satisfactory appearance of the finished book in print and online. After acceptance of the manuscript authors will be asked to prepare the final LaTeX source files (see LaTeX templates online: https://www.springer.com/gb/authors-editors/book-authors-editors/manuscriptpreparation/5636) plus the corresponding pdf- or zipped ps-file. The LaTeX source files are essential for producing the full-text online version of the book, see http://link.springer.com/bookseries/304 for the existing online volumes of LNM). The technical production of a Lecture Notes volume takes approximately 12 weeks. Additional instructions, if necessary, are available on request from lnm@springer.com.

7. Authors receive a total of 30 free copies of their volume and free access to their book on SpringerLink, but no royalties. They are entitled to a discount of 33.3 % on the price of Springer books purchased for their personal use, if ordering directly from Springer.

8. Commitment to publish is made by a *Publishing Agreement*; contributing authors of multiauthor books are requested to sign a *Consent to Publish form*. Springer-Verlag registers the copyright for each volume. Authors are free to reuse material contained in their LNM volumes in later publications: a brief written (or e-mail) request for formal permission is sufficient.

Addresses:
Professor Jean-Michel Morel, CMLA, École Normale Supérieure de Cachan, France
E-mail: moreljeanmichel@gmail.com

Professor Bernard Teissier, Equipe Géométrie et Dynamique,
Institut de Mathématiques de Jussieu – Paris Rive Gauche, Paris, France
E-mail: bernard.teissier@imj-prg.fr

Springer: Ute McCrory, Mathematics, Heidelberg, Germany,
E-mail: lnm@springer.com

Printed in the United States
By Bookmasters

Printed in the United States
By Bookmasters